William J Watkins

Quiz-Compend

A Compend of Human Physiology

William J Watkins

Quiz-Compend
A Compend of Human Physiology

ISBN/EAN: 9783337371500

Printed in Europe, USA, Canada, Australia, Japan

Cover: Foto ©berggeist007 / pixelio.de

More available books at **www.hansebooks.com**

QUIZ-COMPEND.

A COMPEND OF

HUMAN PHYSIOLOGY,

ARRANGED IN THE FORM OF

QUESTIONS AND ANSWERS.

PREPARED AND ESPECIALLY ADAPTED FOR THE USE OF MEDICAL STUDENTS.

By WM. J. WATKINS, M. D.,

GRADUATE OF KENTUCKY MEDICAL COLLEGE, LOUISVILLE, KY.

FIRST EDITION.

LOUISVILLE, KY.:
W. J. WATKINS, M. D., 508 W. Chestnut St.
1891.

—TO—

STUDENTS OF MEDICINE

THIS LITTLE VOLUME

IS RESPECTFULLY CONSIGNED

BY

THE AUTHOR.

PREFACE TO FIRST EDITION.

At the present time, when the student is forced by the rapid progress of medical science to imbibe an amount of knowledge which is far too great to permit of an attempt on his part to master it, any volume which contains the "essentials" of a science in a concise, yet readable form, must, of necessity, be of considerable value. The trite saying that "there is no short road to knowledge" is, of course, as true as it is old; but as long as the effort is made to " cram " the whole science of medicine into a five months' course it is necessary that the hurried student have *condensed books* on the most important branches presenting *essential facts*, so that he need not wade through the more exhaustive and voluminous text-books.

There is no one desires more than the writer that the depth and scope of medical education be increased ; but, in his belief, the evil at present in existence consists in the fact that medical institutions, by granting a degree " too early," make the short road to knowledge the only one which the student with the average amount of cerebral gray matter can possibly travel. The evil lies with the small amount of time required for the obtaining of the degree.

The writer of this little book has, at all times, had in view the interests of the medical student, and should it find a place of average appreciation among them, we will consider that our time and trouble have been justly compensated.

Manuals of this kind are in no way intended to supplant any of the text-books (*or question pamphlets "without answers" now being used in some schools in this country*), but to contain the essence of those facts with which the average student must be familiar, and to arrange the subjects in the form of *questions* and *answers*, from the fact that in reading the standard works the student is often at a loss to discover most important points to be remembered, and is equally puzzled when he attempts to formulate ideas as to the manner in which the *questions* will be put in the *silent* and *unavoidable* "*green room.*"

In producing this little volume we would acknowledge our indebtedness to such standard books as Landois, Dalton, Yeo, etc., but for its foundation and basis throughout we are indebted to Mr. Flint's latest revised edition ; but for a great many of our questions we are under many obligations to our old friend and teacher, Dr. Sam. Cochran.   W. J. W.,

508 W. Chestnut St.

LOUISVILLE, KY., October, 1891.

CONTENTS.

(10)

COMPEND OF HUMAN PHYSIOLOGY.

CHAPTER I.

What is Physiology? *Ans.* Physiology is the study of vital phenomena which are always present in all living things, be they animal or vegetable.

What is Human Physiology? A study of the functions of the human body as exhibited in the healthy state.

How is Physiology divided? Two sub-divisions, known as Animal Physiology and Vegetable Physiology; but it should be remembered that the line of demarkation between Animal and Vegetable Physiology in the lower forms of life is very ill-defined.

From what is the word Physiology derived? From a Greek word Φύσις, "Nature," and λόγος, "a discourse," and in its original meaning was applied to the study of natural history in general, and is really synonymous with the term biology, since it is necessary for the study of either one that vital properties be present in the thing studied.

Define Vegetable and Animal Physiology? (1) Vegetable Physiology treats of the phenomena manifested by which the several structures of the plant is composed. (2) Animal Physiology treats of the phenomena manifested by the organs and tissues of which the animal body is composed.

Which is the most important to physicians? Animal Physiology by far most important.

Describe three most important classes as compared with Human Physiology? (1) "*Nutrition*," which has for its ob-

(11)

ject the preservation of individual—*e. g.*, digestion, absorption, circulation of the blood, respiration, assimilation,
animal heat, secretion and excretion. (2) "*Animal,*" which
bring the individual into a physiological relationship with
external nature—*e. g.*, sensation, motion, language, mental
and moral manifestations. (3) "*Reproductive function,*"
which has for its object the multiplication and preservation
of the species.

**Is there any means by which Human Physiology can be
proven?** Yes. By Anatomy, Chemistry, Pathology, Com-
parative Anatomy, Vivisection, the application of Physics,
etc., etc.

What is the chemical basis of the body? The body may
be studied from a *chemical* or *structural* point of view. Of
the sixty-three elements known to chemists a very small
number, comparatively speaking, are found in any quantity
in living animal matter; however, traces of them all are
frequently present.

What elements found most in abundance? Oxygen, carbon, hydrogen and nitrogen are present in very large proportions in all the tissues, and together compose about
ninety-seven per cent. of the whole body, while sulphur,
phosphorus, chlorine, fluorine, silicon, chloride-sodium, magnesium, calcium and iron are almost indispensable to the
economy, but are widely distributed, and appear in much
smaller quanties.

What two great groups of substances make up the body?
Physiological chemistry teaches us that we have in the
body two sets, or groups of substances known as *nitrogenous* and *non-nitrogenous.*

Which is most important? The nitrogenous perform the
most important functions, and, indeed, form all the active
portions of the organism; as the simplest element of these
nitrogenous bodies may be mentioned *protoplasm.*

What is derived from this? From this, and entering into the formation of it, are albumins, serum-albumins, and by the outcome of still further differentiation we have albuminoids, which is chiefly represented by gelatine, last of those products which, though nitrogenous in character, differ from this in that they are intermediate or effete products of tissue manufacture or waste, as, for example, "*urea*," uric acid, kreatin and kreatinin.

What can you say of non-nitrogenous substances? They consist chiefly of carbohydrates, which contain hydrogen and oxygen in the proportion to form water, as, for example, starch and sugar. Then we have substances containing oxygen in less proportions than the above, namely, *fats*. Chloride sodium occurs all through the tissues, as does also water.

CHAPTER II.

CHEMICAL COMPOSITION OF THE HUMAN BODY.

What do you mean by a chemical composition of the human body? By a chemical analysis the solids and fluids of the body can be first reduced to a number of compound substances, which are termed *proximate principles.* These can be resolved, by an ultimate analysis, into fifteen chemical elements. The different chemical elements thus obtained, and the proportion in which they exist, are shown in the following table : *

*a*Oygen	72.00	O H and C are found in all the tissues	
*b*Hydrogen	9.10	and fluids of the body without exception.	
*c*Nitrogen	2.50	O H C and N found in most of the	
*d*Carbon	13.50	fluids of the tissues except *fats.*	
*e*Sulphur	14.7	In fibrin, casein, albumin, gelatine, as potassium sulpho-cyanide in saliva ; as alkaline-sulphate in *urine* and *sweat.*	
f Phosphorus	1.15	In fibrin and albumin ; in brain, as trisodium ; phosphate in blood and saliva, etc.	
*g*Calcium	1.30	As calcium phosphate in lymph, chyle, blood, saliva, bones and teeth.	
*h*Sodium	.10	As chloride sodium in all the fluids and solids of the body except enamel ; as sodium sulphate and phosphate in blood and muscles.	
*i*Potassium	.026	As potassium chloride in muscles ; generally found with sodium, as sulphates and phosphates.	
*j*Magnesium	.001	Generally in association with calcium, as phosphates in bones.	
*k*Chlorine	.085	In combination with sodium, potassium and other bases, in all the fluids and solids.	
*l*Fluorine	.08	As calcium fluoride in bones, teeth and urine.	
*m*Iron	.01	In blood globules, as peroxide in muscles.	
*n*Silicon	traces	In blood, bones and hair.	
*o*Manganesium	traces	Probably in hair, bones and nails.	

*From Burbaker.

Summary. Of the four chief elements which combined make up *97 per cent.* of the body, O H N are eminently *mobile, elastic,* and possesses great *atomic heat.* C H N are distinguished for the narrow range and feebleness of their affinities and chemical inertia. C has the greatest atomic cohesion. O is noted for the number and intensity of its combinations, and its remarkable display of chemical affinity, combining with every element except fluorine.

With the exception of the gases O H and N, what can you say? Other elements do not exist alone in the body, but are combined in characteristic proportions to form compounds, the *proximate principles*, the ultimate compounds to which the fluids and solids can be reduced.

How do proximate principles exist in the body? Under their own form and can be extracted without losing their distinctive properties.

How are proximate principles divided? There are about *one hundred* proximate principles, which are divided into *four* classes, viz : *inorganic, organic non-nitrogenized, organic nitrogenized* and *principles of waste.*

1. INORGANIC PROXIMATE PRINCIPLES.

Although this subject is treated elsewhere, give a brief outline of those four classes of proximate principles? 1. Inorganic proximate principles are oxygen, found in lungs and blood. Hydrogen, in stomach and intestines. Nitrogen, in blood and intestines. Carbonic anhydride, found in expired air of lungs. Carburetted hydrogen and sulphuretted hydrogen found in lungs and intestines. Water, in all the solids and fluids. Sodium chloride, found in all the solids and fluids except the enamel of the teeth. Potassium chloride, in muscles, liver, saliva, gastric juice, etc. Ammonium chloride, gastric juice, saliva, tears, urine. Calcium chloride, bones, teeth and urine. Calcium carbonate, bones, teeth, cartilage, internal ear and blood. Calcium phosphate, magnesium phosphate, sodium phosphate, potassium phosphate, found in all fluids and solids of the body. Sodium sulphate, potassium sulphate, universal except milk, bile, and gastric juice. Sodium carbonate, potassium carbonate, bones, blood, urine, lymph, etc.

Of this class which most important? *Oxygen* is one of the constituents of all the fluids and solids of the body ; it is found in a free state in the respiratory passage and intestinal tract. The function of O in the body appears to be the oxidation of the albuminous, oleaginous, and saccharine compounds to their ultimate forms, urea, carbonic acid, water, etc.

As to whether this is brought about by direct oxidation or by a fermentative process is yet unknown. As O only enters into combination under a high temperature, it is assumed that it exists in the body under the form of ozone, OO^3, which possesses remarkably active oxidizing powers.

What can you say of hydrogen? It is also a constituent element of almost all the compounds of the body ; existing in a free state in the intestinal tract, when it is produced by a decomposition of organic substances.

What is the function of hydrogen? Its function in the system is unknown, yet it is asserted by Hoppe-Seyler that hydrogen unites with neutral oxygen, O_2, in the tissues, forming water and liberating oxygen in the nascent state, which becomes the oxidizing agent. The process is represented in the following equation : $H H + O_2 + n = H_2 O + O n$, in which n represents the oxidizing substance.

To combine O_2 and H what have you? The combination form *water*, which is the most essential constituent of the body.

How much of the body is composed of water? About seventy per cent. of the entire body weight.

Its most important function? A vehicle, lubricant, solvent, etc.

How introduced into the body? In the form of drink, and as a constituent of all kinds of food. The average quantity consumed daily is about four pints.

How does water act as a solvent, etc.? Gives general pliability to various tissues, and promotes the passage of inorganic and organic matter through animal membranes. It also promotes chemical changes which are essential to absorption and assimilation of food and the elimination of products of waste.

How eliminated? By the skin, lungs and kidneys.

What can you say of sodium chloride? It is present in all the solids and fluids of the body, with the exception of the enamel of the teeth, regulates osmotic action, holds the albuminous principles of the blood in solution, and preserves the form and consistence of blood corpuscles and the cellular elements of the tissues by regulating the amount of water entering into their composition.

How is chloride sodium introduced? In the form of nearly all foods and drinks, each person containing four or five ounces; of this, about ten per cent. is consumed and eliminated daily—thrown off chiefly by skin and kidneys.

What part does calcium-phosphate play in the body? Most abundant of all the inorganic principles, with the exception of water, and is present to a great extent in bone, teeth, muscles and milk. It gives the requisite consistency and solidity to the different organs, and in the blood it is held in solution by the albuminous constituents. The *sodium* and *potassium phosphates* are present in most of the solids and fluids, and give to them their alkaline reaction. They are chiefly derived from the food.

2

2. ORGANIC NON-NITROGENIZED PRINCIPLES.

From what are Organic Non-nitrogenized Principles derived? Derived mainly from the vegetable world, but are also produced within animal body.

Into how many classes can you define this second class of proximate principles? Into four subdivisions. (1) The *carbo-hydrates*, comprising *starch* and *sugar*, bodies in which the O and H exist in proportion to form water; (2) The *fats*, bodies having the same elements entering into their composition, but with the carbon and hydrogen increased and the oxygen diminished in amount; (3) *Fatty acids*, which combine with sodium, potassium, and calcium, are found as salts in various fluids of the body, such as blood, chyle, fæces, etc. (4) *Alcohols.* In this group we have Glycerine, Cholesterin, and Alcohol. *Glycerine* is chemically a triatomic alcohol in combination with the neutral fats of the body. During pancreatic digestion it is set free. This by many physiologists is supposed to be directly concerned in the production of glycogen. *Cholesterin* is a crystallizable substance largely present in the bile, though it is found in other fluids and solids. *Alcohol* has been found in the *urine*. It is supposed to be the result of an alcoholic fermentation in the intestines.

3. ORGANIC NITROGENIZED PRINCIPLES.

Give a brief outline of this third class? This class are organic in their origin, being derived from the animal and vegetable world. They are taken into the body as food, and are appropriated by the tissues, and constitute their organic basis. They differ from the *non-nitrogenized* substances in not being crystalline, but amorphous, in having a more complex but just as definite composition, and containing in addition to C H O, nitrogen, with at times, sulphur and phosphorus.

The proteids possess peculiarities which distinguish them from all other substances, viz., a *molecular mobility*, which permits isomeric modifications to take place with great facility; a *catalytic influence*, in virtue of which they promote, under favorable conditions, chemical changes in other substances; *c. g.*, during digestion salivin and pepsin cause starch and albumin to be transformed into sugar and albuminose respectively.

Briefly outline the principal nitrogenized (or proteid compounds) bodies as found in the organism? They are quite numerous, and although they resemble each other in many particulars, yet there are important distinctions, and can be better understood in the following groups:

1. NATIVE ALBUMINS.—Proteid bodies soluble in *water*, many *acids*, and usually in *alkalies;* coagulable at a temperature of from 140° to 163° F.

a. Serum Albumin, the principal form of albumin found in the animal fluids and solids.

b. Egg Albumin, not found in ordinary tissues, but present in white of egg.

2. GLOBULINS.—Proteid bodies insoluble in water, but soluble in solutions of sodium chloride.

a. Globulin, found in many tissues, but largely present in crystalline lens.

b. Myosin, found in the muscles in life, in a fluid condition; after death it undergoes coagulation, giving rise to the rigidity of the muscles.

c. Paraglobulin, present in blood and obtained from it by passing a stream of *carbon dioxide* through it; it is also precipitated by adding *sodium chloride.*

d. Fibrinogin, present in serous fluids and blood, and can be precipitated by the prolonged use of carbon dioxide; it is also precipitated by the addition of 12 or 16 per cent. of sodium chloride.

3. DERIVED ALBUMINS.—Proteid bodies which are not coagulable by heat; insoluble in pure water and in salt solutions; soluble in both acid and alkaline solutions.

a. Acid Albumin, found principally in stomachs during the first stage of digestion, the result of the action of the hydrochloric acid upon the albumin of the food.

b. Alkali Albumin, found in the intestine during pancreatic digestion, the result of the action of alkalies upon the albumin of the food.

c. Casein, the chief proteid of milk; it is precipitated by acetic acid and rennet.

4. PEPTONES.—These bodies are found in the stomach and intestinal tract by the action of the gastric and pancreatic juices upon the albumins of the food. Soluble in water, alkaline and acid solutions; non-coagulable by heat; very diffusible. They are precipitated by tannic acid and alcohol.

5. ALBUMINOIDS.—The albuminoids are the result of various modifications of albumins occurring during the nutritive process, as well as by the action of various external influences.

a. Mucin, the characteristic ingredient of mucus, secreted by the mucous membrane, giving to it its viscidity.

b. Chondrin, found in cartilage.

c. Gelatin, found in connective tissue, tendons, ligaments, bones, etc.

d. Elastin, found in elastic tissue.

e. Keratin, found in skin and epidermic appendages, nails, hair, horn, etc.

6. FIBRIN.—A filamentous albumin obtained by washing blood clots. It is insoluble in water and mineral acids.

4. FOURTH CLASS OF PROXIMATE PRINCIPLES.

Under what head is this fourth class generally known? Principles of waste.

Known by what names? Urates.

Name the principal elements of " Waste ? " The principal elements of Waste are:

Urea,	Xanthin,	Sodium,	⎫
Creatin,	Tyrosin,	Potassium,	Urates.
Creatinin,	Hippuric Acid,	Ammonium,	
Cholesterin,	Calcium Oxalate,	Calcium,	⎭

Those principles are represented as what ? Waste of organic origin, arising within the body as products of disassimilation or retrograde metamorphosis of the tissues. They are absorbed by the blood, carried to the various excretory organs, and by them eliminated from the body. The excrementitious substances will be fully considered under Excretion.

CHAPTER III.

STRUCTURAL COMPOSITION OF THE BODY.

Give the proximate quantity of the chemical elements and proximate principles of the body weighing 154 lbs.? *Ans.* This has been estimated as follows :

	Lbs.	Oz.		Lbs.	Oz.
Oxygen	111	. .	Water	111	. .
Hydrogen	14	. .	Albuminoids	23	7
Nitrogen	3	8	Fats	12	. .
Carbon	21	. .	Calcium Phosphate . .	5	13
Calcium	2	. .	" Carbonate . .	1	.
Phosphorus	1	12	" Fluoride	. .	3
Sodium, etc.	.	12	Sodium Sulphate, etc. . . .	. .	9
	154	. .		154	. .

CHAPTER IV.

THE BLOOD.

What is the most abundant and highly organized fluid of body? *Ans.* Blood.

What does it contain? All the elements necessary for nutrition.

What does it receive? Products of waste.

What do these processes require? Regeneration of the blood and elimination of effete matter.

What is extra vascular tissue? That not directly supplied with blood.

How nourished? By imbibition.

What tissues are supplied by blood? Those in which nutrition is active.

Symptoms produced by abstracting large quantity of blood? Stops the vital powers.

Effect of returning the blood? Suddenly revives.

Why does death not occur immediately where large artery is cut? The hemorrhage weakens the heart.

What is transfusion? Injecting blood of one person into another.

Part inject? Red corpuscles.

Why? Vivifying power lies in them.

Amount? Three to six fluid ounces.

Dangers of transfusion? Shock to heart and danger of getting air into veins.

Danger if air gets into veins? It can't circulate through the pulmonary capillaries and causes death.

Danger if it gets into arteries? None.

What is blood? Fluid found in the blood vessels.

Amount? One-tenth weight of the body.

How determined? Decapitate, drain the blood, then wash blood vessels with water, evaporate this to a clot and compare to other clot to see how much blood it would represent.

When greatest? During or just after digestion.

More blood in male or female? Male.

At what period of life is there least blood, proportionately? Infancy.

What two elements make up blood? Plasma and corpuscles.

State the characters of blood? Red, opaque fluid, saline taste, alkaline in reaction, characteristic odor.

Why opaque? Due to the difference of refractive power between corpuscles and plasma.

Taste? Saline.

Why? Due to chloride of sodium.

Odor? Characteristic of the animal.

How brought out? Add a mineral acid.

Reaction? Alkaline.

Why? Due to the basic phosphates and carbonates.

Specific gravity? 1052 to 1057.

Greatest, male or female? Male.

Color in arteries? Bright red.

Color in veins? Blue.

When red in veins? During the period of gland activity.

Why? Circulates too fast to give up its oxygen.

Warmest which side of the heart? Right.

Why? Has not been cooled in going through the lungs as that of left side.

Warmest, arteries or veins? Arteries.

Hottest blood in the body? That of the hepatic vein.

Temperature? One hundred and seven degrees Fahrenheit.

Warmest in portal vein or abdominal aorta? Portal vein.

Temperature of the blood? Ninety-eight and one-half degrees.

Anatomical elements of the blood? Red and white corpuscles.

Define "anatomical element?" Any thing that undergoes waste and repair.

Proportion of red corpuscles to blood? Little less than one half.

More in male or female? Male.

Shape of red corpuscle? Biconcave circular disc.

Where thickest? At the edge of vessels.

Consistence of red corpuscles? Consistence of liver.

Why think they are elastic? On account of the persistency with which they keep their shape.

If blood does not coagulate, why do the red corpuscles go to the bottom? Due to their high specific gravity.

What is their specific gravity? 1088 to 1105.

What is that of plasma? 1028.

Appearance of red corpuscles under the microscope? Nucleated.

Why? Can't focus the edge with the center.

Color of red corpuscles? Amber.

Illustrate? Dilute it with water.

What do you mean by Rouleaux? Arrangement of the red corpuscles like piles of coin.

Cause of this arrangement? An adhesive exudation.

Size of red corpuscle? One thirty-five hundreths of an inch.

Largest in which animal? Elephant and sloth.

Different shape in what animals? Camel and llama.

How tell blood of man from lower animals? By their size.

Relation of muscular activity to the size of the corpuscle? Inverse.

Peculiarity of red corpuscles of invertebratas? Are oval and have a nucleus.

How count red corpuscles? Dilute blood and put in glass tube.

Number to one twenty-fifth of an inch? Four million.

In what vessels most abundant? Veins.

In what vein are they most abundant? Splenic.

Least abundant? Hepatic.

What does this indicate? That they are manufactured in the spleen and destroyed in the liver.

Post mortem appearance? Shriveled and serrated.

How make them resume their original form? Add fluid with the same density as plasma.

What is the structure of red corpuscles? Homogeneous.

Why do you think so? A membrane has never been demonstrated.

Name the first blood vessel? Area vasculosa.

When does the blood become red? When the fœtus is one-tenth inch in diameter.

When does corpuscle appear first in fœtus? About the same time.

Relative size at first? Thirty to one hundred per cent. larger than in adult.

Shape then? Most circular ; some oval or globular.

Appearance red blood corpuscles under microscope? Nucleated.

How long do they remain so? Three or four months.

Give reason for thinking that they are not produced by any special organ? They exist before any organ.

Why think they are not formed from leucocytes? They exist before the leucocytes.

Origin of first corpuscles? They are generated *de novo*.

In adult what is their origin? Spleen, lymphatic gland and red marrow of the bones.

Color of marrow of bones in fœtus? Red.

Name? Lymphoid.

Color of marrow in long bones of adult? Yellow.
Where find red? Short bones.
Part spleen plays? Manufactures red corpuscles.
Part liver plays? Destroys them.
What becomes of coloring matter? Goes to form other pigmentary matter.
Relation of anæmia to corpuscle forming function of the marrow? It increases it.
Function of red corpuscles? Carry oxygen.
What part of blood most necessary to life? Red corpuscles.
Symptoms when all the arteries to a part are tied? Loss of function.
What is the oxygen absorbing capacity of the blood? Ten to thirteen times as much as water.
On what does it depend? On red corpuscles.
Other name for the white corpuscles? Leucocytes.
Why have they this name in blood? Because they are found elsewhere.
Where are leucocytes normally found? Blood, lymph, chyle, cholestrum and vitreous humor.
Where pathologically found? Pus and irritated mucous membranes.
Shape? Globular.
In what part of vessel does red corpuscle circulate? Middle.
Where do the white circulate? Along the sides of vessels.
Cause? Their adhesive character.
Size of leucocytes? One-twenty-five hundredth inch.
In what fluid do you find them in greatest abundance? Pus.
What shape may they assume? Polygonal or ovoid.
When do they increase in size? When confined in tissue.

Number as compared with red corpuscles? One to seven hundred and fifty or one thousand red.

Specific gravity? Ten per cent.

When do they first appear? In early fœtal life.

Why think spleen has to do with their production? Find more in splenic vein than splenic artery.

Why in lymphatic glands? More in blood going to than coming from the glands.

What is leucocythemia? Abnormal amount of leucocytes in the blood.

What does leucocyte mean? White body.

Origin of leucocytes? Spleen, marrow of the bones, and lymphatic glands.

Anatomical changes in the spleen and lymphatic glands in leucocythemia? They become enlarged.

Functions of the leucocytes? Thought to aid in coagulation of blood.

Other name for blood plaques? Elementary corpuscles.

Shape? Flat circular discs.

Structure? Homogeneous.

Size? One-sixth that of the red corpuscle.

Color? Grayish tint.

Number? One to twenty or thirty red corpuscles.

Where are they distinct? In the circulating blood.

When do they adhere? When the blood is drawn.

Changes they undergo out of the body? They become ovoid or elongated.

Function of the plaques? To form red corpuscles.

Composition of red corpuscles? Globulin, inorganic salts and coloring matter.

Rich in what inorganic salts? Salts of potassa.

What is globulin? Organic nitrogenous proximate principle of the red corpuscle.

How extracted? By adding chloride of sodium to defibrinated blood.

What is the coloring matter of red corpuscle? Haema-globin.

How obtained? By percolating blood with ether.

Shape of the crystals? Four sided prisms or rhomboids.

Color? Purplish.

How is it precipitated? By acetic acid, nitrate of mercury, chlorine gas and ferrocyanide of potassium.

In what is it soluble? Water and weak alkaline solutions.

Proportion of coloring matter in blood? One hundred and twenty-seven parts per one thousand.

In dried corpuscles? Twelve-thirteenths to nine-tenths of the dried corpuscle.

With what does coloring matter unite? With oxygen.

What kind of union? Unstaple.

What is plasma? Watery element of the blood.

Specific gravity of it? 1028.

In studying the analysis of the blood how is it divided? Into proximate principle or chemical constituent.

Define chemical constituent? Something that exists as such in the body, and when extracted and decomposed changes properties.

Is it an element or compound? May be either.

Define an element? Simple and undecomposable substance.

Name one? Oxygen.

Define a compound? Substance made up of two or more elements.

Name one? Water.

What is the first class of chemical constituents? Inorganic.

What are the properties of the inorganic? Inorganic in origin, definite chemical composition, crystallizable, need no preparation for introduction, undergo no change in the organism, and always exist in combination with organic.

When is there a deficiency of elimination? Old age.

What does it cause? Calcareous degeneration.

Most important of the inorganic? Water.

Where do you find it? Throughout the system.

How do you know when you need it? Get thirsty.

Where thirst referred? Throat and fauces.

Where exist? In the tissues.

Amount of chloride of sodium, or common salt, in the blood? Four parts per 1000, or about five and one-half ounces in each person.

Function of chloride of sodium? It gives density to the plasma, aids in osmosis and determines transudation.

Amount of inorganic salts? Six to eight parts per 1000.

Function of them? Prevent solution of the corpuscle.

Which of them are decomposed in the system? Carbonates and phosphates.

Which are formed in the system? Organic salines.

Function of pneumic acid? To decompose the carbonates and bicarbonates.

Origin of pneumic acid? Formed in the organism.

Origin of lactic acid? From decomposition of lactose.

Function? Same as pneumic acid.

Name the organic salines? Lactates and pneumates.

Origin? Action of lactic and pneumic acid on the bases.

Name the organic non-nitrogenized constituents of the blood? Olein, margarin, steirine, lecithin, glucose, glycogenic matter and inosit.

Why is lecithin placed in this group? It has many of the properties of the fats.

Where found? Blood, bile, nervous tissue, and yolk of egg.

What organ first produces sugar? Placenta.

After birth in what blood find it? Blood going from liver to lungs..

What becomes of it? It is destroyed in the lungs.

Is it ever deposited? It is not.

Is it ever eliminated? Only where large quantity is ingested.

Function of sugar? Thought to produce fat.

How are fats and sugars distinguished from the others? They contain carbon, oxygen and hydrogen.

How are fats introduced? Form of adipose tissue.

Where digested and by what? In small gut by the pancreatic juice.

By what is it absorbed? Lacteals.

Destination of fats? Either destroyed or deposited.

Ever eliminated? No.

Function? Thought to keep up animal heat.

Name the organic nitrogenized principles of blood. Plasmine, serine, peptones and coloring matter.

Properties of this class? Organic in origin, have no definite chemical composition, are not crystallizable, need preparation for introduction and undergo change in the organism.

What is plasma? Watery element of blood.

What is plasmine? Organic nitrogenous proximate principle of blood.

Amount? Twenty-five parts per one thousand.

How obtained? By adding saturated solution of chloride of sodium to plasma.

How decompose it? Add water and whip with twigs.

What are the products of its decomposition? Metalbumin and fibrin.

How many parts of each? Metalbumin twenty-two parts, fibrin three parts.

Is fibrin a proximate principle? No.

What is it? A product of decomposition of plasmine.

What is metalbumin? Also a product of decomposition of plasmine.

What is serine? Organic nitrogenous proximate principle of blood.

How does it differ from albumin? It is more osmotic and is not coagulated by ether.

Amount of serine? Fifty-three parts per one thousand.

What represents the albumin of the older writers? Met-albumin and serine.

Does the blood contain albumin? It does *not*.

What are peptones? Organic nitrogenous proximate principles of blood.

Origin? Products of stomachic digestion.

Has the coloring matter ever been isolated? No.

What are paraglobine and fibrinogen? Fibrin producing principles.

In what form is blood in the vessels? Fluid.

How if circulation is disturbed? Clots.

Technical name for a clot? Crassamentum.

What blood coagulates most rapidly? That of birds.

Why will arterial blood coagulate more rapidly than venous? There is more fibrin formed.

What is the first thing you notice in the process of coagulation? A thin pellicle forms on top.

In what time? One to six minutes.

Next thing? It gets gelatinous on the sides.

In what time? Two to seven minutes.

Next thing? Entire mass becomes gelatinous.

In what time? Seven to sixteen minutes.

What becomes of the clot? It separates into two parts.

Cause? Contraction of the fibrin.

In what time? Ten to twelve hours.

What does the solid part contain? Corpuscles and fibrin.

What does the serum contain? The watery part.

Amounts of each? About equal.

On what does the consistence of clot depend? On the rapidity of coagulation.

Color of a venous clot? Blue.

Changes when exposed to air? Becomes red at the edge.

Why? It takes on oxygen.

What do you mean by "buffy coat?" Yellow layer on top of the clot.

Cause of the yellow color? A peculiar coloring matter which has never been isolated.

Cause of buffy coat? Slow coagulation.

Old significance of it? Thought it was pathognomonic of inflammation.

What is serum? Watery part of the clot.

What is plasma? Watery part of the blood.

Which is physiological? Plasma.

Color of serum? Straw color.

Specific gravity? 1028.

To what is the coagulation of blood due? Formation of fibrin.

What expresses the serum? Contraction of the fibrin.

When do the corpuscles remain in the clot? Due to the adhesive matter which covers them.

What kind of stream does blood coagulate more rapidly? Slow and small stream.

What kind of vessel would hasten coagulation? A rough one.

How can you produce instantaneous coagulation? Receive the blood on a cloth.

Influence of air on coagulation? It favors it.

Influence of a vacuum? Increases it.

Effect of freezing the blood? Retards coagulation while it is frozen.

What chemicals will prevent coagulation? Soda and potash salts.

Why does menstrual flow not coagulate? Vaginal secretion keeps it fluid.

Does blood coagulate in vessels after death? Yes.

In what time? Twelve to twenty-four hours.

Chiefly in what vessels? The veins.

Why are arteries empty? They contract after death.

Coagula are largest on which side of the heart? Right side.

Does coagula form during life? Yes.

What were they called by the old writers? Polypi.

How do they occur? When death is gradual.

How can you tell anti from a post-mortem clot? Antemortem is firmer, whiter, and more closely adherent.

Effect of projections in the vessel? Favors their formation.

Name of stationary clot? Thrombus.

Of a floating one? Embolus.

Condition the blood assumes when effused in tissue or cavities? It coagulates.

In what cavity will it remain fluid? Tunica vaginalis.

What is hæmorrhagic diathesis? A tendency to bleed.

Cause? Lack of fibrin-producing principle.

Why will a cut vessel bleed more than a torn one? The torn one exposes more blood to the air, favoring coagulation.

Office of coagulation? To arrest hæmorrhage.

Cause of coagulation? Decomposition of plasmine by a ferment.

Origin of the ferment? Due to changes in the leucocytes.

Do the red corpuscles play any part in coagulation? No.

What is paraglobin? About the same as metalbumin.

Peculiarity of the blood of a leech bite? It bleeds considerably and remains fluid.

What blood does not coagulate very readily after death? That of renal and hepatic veins and capillary blood.

Is fibrin a proximate principle? No ; it is a product of decomposition of plasmine.

CHAPTER V.

THE HEART.

Where do you find it? *Ans.* Thoracic cavity in middle mediastinum.

Peculiarities of its fibres? Small, striated and anastomose freely, have no sarcolemma, and contain granular matter.

Give arrangement of muscular fibres? Two layers ; the external is oblique and the internal circular.

Origin of external? At base of heart.

Where become spiral? At apex.

Where go? To form the columna carneæ.

Describe the different kinds? (1) Columna carneæ proper, simply fleshy projections in the heart ; (2) Columna papillaria, give attachment to the cordæ tendinæ ; (3) trabeculæ carnæ, are attached at each end and free in the middle.

Thickness of right auricle? One line.

Left auricle? One and one-half lines.

Right ventricle? Two and one-half lines.

Left ventricle? Seven lines.

What is systole? Contraction of the heart.

What is diastole? Dilatation of the heart.

To what part are these terms referred? To the ventricles.

Why? They are most important to the circulation.

When are the auricles not receiving blood? During auricular systole.

From auricles where does the blood go? To ventricles.

Effect this has on the ventricles? Dilates them.

What follows? The ventricles contract.

From right ventricle where does the blood go? Through pulmonary artery to lungs.

From left ventricle where does it go? Through the aorta to general system.

What are the movements of the heart? Forward, upward and from left to right.

Why does it raise up? Because the anterior fibres are longer than the posterior.

Why come forward? Due to distention of the great vessels.

Why go from left to right? Because the fibres run from right to left.

What part of the heart is fixed? Base.

What is the origin of a muscle? A fixed point.

Why does the heart twist? On account of the spiral course of the superficial fibres.

Why does it get hard? Due to muscular contraction.

Does it shorten or elongate? It shortens.

What was the old idea? That it elongated.

How was it determined? Pin the base to a board and measure it.

What is the impulse of the heart? Beat against the chest wall.

When does it occur? During systole.

Proof? If you introduce your finger into the chest you feel it at the same time as pulse.

What represents time? One.

Time of auricular systole? One-fifth.

How was it determined? By means of the cardio-graph.

Describe the contraction. Feeble contraction followed immediately by relaxation.

Time of ventricular systole? Two-fifths.

Describe it. Powerful contraction followed by sudden relaxation.

What time do the ventricles rest in a day? Three-fifths.

Time auricles rest? Four-fifths.

Time of ventricular systole? Two-fifths.

What is the force of the heart? Fifty-one and one half pounds.

How determined? Multiply area of left ventricle by arterial pressure.

Are there any valves in the right auricle? Yes.

Describe them. Coronary guards, opening of coronary vein and Eustachian extend between foramen ovale and the auriculo-ventricular opening.

Do they assist in circulation? Not in an adult.

What causes the blood to go from right auricle to ventricles? Force from behind and opening in front, and contraction of the auricle.

Functions of auriculo-ventricular valves? To prevent regurgitation into auricle during systole.

When are they closed? During systole.

What is the heart doing then? Emptying its contents.

What valves are then open? Semilunar.

Function of the semilunar valves? Prevent reflux of blood into ventricles.

When are they closed? During diastole.

What is the heart then doing? Filling.

State the appreciable phenomena of the heart? Pulse, impulse and sound.

In studying the rhythm of heart, where begin? Auricular systole.

Why? There is where the blood first gets in.

Studying rhythm of sounds, where begin? Ventricular systole.

Why? It is heard first.

Name the sounds of the heart? First and second.

Why call the long one the first sound? It is heard first.

Which sound is simple? Second.

Cause of second sound? Closure of the semilunar valves.

How prove it? Tie them back and you will not hear the sound.

What elements make up the first sound? Valvular, prolonged and booming.

What is the cause of the booming element? Beat against the chest wall.

Cause of prolonged? Contraction of the heart muscle.

Cause of valvular? Closure of the auriculo-ventricular valves.

How make the first sound like the second? Put the patient on his back and towel on chest.

What are the causes of the first sound? Beat against the chest wall, muscular contraction and closure of the auriculo-ventricular valve.

Where heard loudest? At the apex and left side.

Why? Transmitted to chest by apex beat.

Condition of the heart then? Emptying.

Course of the blood during the first sound? Into the arteries.

What keeps the blood from going back into auricles? The auriculo-ventricular valves.

On which side of the heart may the valves be diseased? Left.

Abnormal sounds caused by disease of the mitral valves are heard where? Apex and left side.

When? Just before or during systole.

Condition of ventricle just before hear first sound? It is filled.

, **Blood flowing from auricle into what?** Ventricle.

Does it usually cause a sound? No.

Why not? The opening is sufficiently large for blood to pass without transmitting noise.

What would obstruct the flow from auricle to ventricle? Where valves are tied and can't open.

Would it then cause a sound? Yes.

Why? Blood is being forced through small opening.

Where do you hear it? Apex and left side.

When? Just before systole.

What is the heart doing? Filling.

Cause of this sound? Blood being forced through a small opening.

What is it called? Presystolic.

Why? Because it is heard just before systole.

The condition of the valves is called what? Mitral stenosis.

Why? Because the opening is made small.

What is the condition of the heart during the first sound? Emptying.

Course of the blood? Into the arteries.

If mitral valves do not close sufficiently what is the direction of the blood? Both directions.

Effect of this on first sound? Masks it.

This condition of valves is called what? Mitral insufficiency.

What is the sound called? Mitral regurgitation.

Where heard? Apex and left side.

When? When you feel the pulse?

Direction of the blood when you hear it? Into arteries.

If heart is beating rapidly, how tell the second from the first sound? By the position you hear it.

If you hear an abnormal sound at apex just before you feel the pulse, what is it called? Presystolic.

Cause of it? Mitral stenosis.

When you feel the pulse? Mitral regurgitation.

Name? A systolic brew.

Cause? Blood flowing back into auricle.

Where do you hear the second sound loudest? Base and right side.

Why? It is conducted along the great vessels.

Cause of the second sound? Closure of semilunar valves.

Does blood cause sound normally in passing through the semilunar valves? No.

When would it do so? Where valves tied and can't open.

Why? Blood is forced through small opening.

Where do you hear it? Base and right side.

When? As you feel the pulse.

What is it called? Aortic stenosis.

Condition of heart when you hear it? Emptying.

Function of the semilunar valves? To guard the arteries.

If they don't close the opening, where does the blood go? Back into ventricle.

Effect of this on the second sound? Masks it.

What would you have? An abnormal sound.

Where do you hear it? Base and right side.

When? When you don't feel the pulse.

What is it called? Aortic regurgitation.

Why? Because the blood is flowing back.

Condition of the heart when you hear it? Filling.

Frequency of the heart's action? Seventy-two beats per minute.

Period of life most frequent? At birth.

Frequency at that time? One hundred and twenty-six to one hundred and forty per minute.

How frequent in adult male? Seventy.

Adult female? Seventy-six to seventy-eight.

Effect of old age? Increases it.

Relation of heart's action to respiration? Direct relation.

Kind of pulse have in asphyxia? Slow and strong.

Why? Gets so to overcome the obstruction in the systemic capillaries.

How is arterial pressure then? Very high.

If the obstruction is not overcome, how will the heart die? Distention of the heart and paralysis of its muscular fibres.

Name the different kinds of muscular fibres. Striated and non-striated.

To what class does the heart belong? Striated.

How differ from the other striated? It is involuntary.

Things necessary for striated muscle to contract? Must receive a stimulus.

What is the normal stimulus? Nerve force.

If nerve is cut, what effect has it on muscle? Muscle becomes paralyzed.

How does the heart differ from other muscles? It needs no stimulus.

How is contraction of heart influenced by circulation of blood through it? Causes regular and powerful contraction.

Give illustration. Cut out the heart and pass liquid of same density as blood through it, it will contract.

On what does the intermittent action and successive action of the heart depend? Its contained sympathetic ganglia.

Where situated? In the heart substance at the base.

Proof? Remove the heart and cut it transversely, and the apex will cease to beat.

Where find the ganglion of Remak? At expansion of the inferior vena cava.

Ganglion of Bidder? Between the left auricle and ventricle.

Function of these two? Motor.

Where find ganglion of Ludwig? Between the two auricles.

Function? Inhibitory ganglion.

Where does contraction of the heart start? At the ganglia.

What part of the heart has the power of intermittent and regular contraction? Muscular fibres of auricles and upper third of ventricles.

Which surface of heart is most excitable? Interior.

What is the natural stimulus to the heart? Passage of blood through its cavity.

How is heart's action influenced by destruction of the cardiac plexus? Heart's action is immediately arrested.

Effect on heart by cutting the sympathetic? Lessens the cardiac movements.

Stimulating it? Increases its movements.

Name the inhibitory nerve. Pneumogastric.

Effect on heart if cut? Beats faster.

If stimulated? Beats slower.

From what nerve does this action come? Spinal accessory.

Action of digitalis on the heart? Slows and strengthens it.

How? It stimulates the pneumogastric nerve.

Does it influence the heart if the pneumogastric is cut? No.

Is the action of the pneumogastric nerve direct or reflex? Direct.

What do you mean by direct? Where the force originates in the center.

What mean by reflex? Force sent back.

Does reflex inhibition ever occur? Yes.

Give illustration? Tap the intestines of a frog and his heart stops.

Cause of syncope? Stimulus sent to the inhibitory center in the medulla and then back to the heart.

Treatment? Swing the patient by the feet to stimulate the center.

CHAPTER VI.

FOETAL CIRCULATION.

Where do you begin to trace it? Placenta.

Through what vessel does it get to the child? Umbilical vein.

Carries what kind of blood? Arterial.

Where does it enter the abdomen of child? At the umbilicus.

It goes to what organ? Liver.   •

What vein does it meet? Portal vein.

Greater part is distributed to what? Liver.

Does it all pass to the liver? No.

Blood that does not, goes where? Into inferior vena cava.

How? Through the ductus venosis.

Where does the inferior vena cava empty? Right auricle.

What part of the right auricle? Posterior and inferior aspect.

The current is directed toward what opening? Foramen ovale.   .

What keeps the blood from going into right ventricle? Eustachian valve.

Give the attachment of the Eustachian valve. From the foramen ovale to the auriculo-ventricular opening.

When is it largest? At birth.

How does blood get from right to left auricle? Through the foramen ovale.

Where does it go from left auricle? To left ventricle.

Where does it go from left ventricle? Into aorta.

From aorta? To the tissues.

Does it all pass to the upper extremity? No.

How is the blood returned from the upper extremity? By the superior vena cava.

How does this empty into right auricle? Upper and posterior part.

Direction of the current? Toward the auriculo-ventricular opening.

Where does blood go from right auricle? To right ventricle.

From right ventricle? Pulmonary artery.

From pulmonary artery? To lungs and to aorta.

How much passes to the lungs? Enough for nutrition.

Where do you find the ductus arteriosus? Between the pulmonary artery and aorta.

How does the blood get back to the placenta? Through the umbilical arteries.

Peculiarity of the internal iliac arteries in the fœtus? Larger than in adult, and give off the umbilical.

How many cavities in the fœtal heart? Two.

How are the auricles connected? By foramen ovale.

How are the ventricles connected? By the ductus arteriosus.

Peculiarity of fœtal heart? Foramen ovale and Eustachian valve.

Peculiarity of fœtal arteries? Have the ductus arteriosus and umbilical.

Of fœtal veins? Ductus venosis and umbilical vein.

Changes after birth? Arteries and veins become obliterated, foramen ovale closes and the Eustachian valve atrophies.

Why is the liver so large at birth? It gets a large amount of arterial blood.

Why is the head so large? It also gets large amount.

Why are the lower extremities so small? Get very little arterial blood.

From which side does the membrane grow that closes the foramen ovale? Left.

When? Tenth day after birth.

CHAPTER VII.

ARTERIES.

How many are there? Two.

Name them. Aorta and pulmonary.

Why are they called arteries? In olden times they were thought to contain air.

Why thought to contain air? They are found empty and open after death.

Why are they found open? Their thick walls hold them open.

How does the pulmonary artery differ from the aorta in its anatomy? Its walls are thinner and more distensible.

What is the course of arteries? Direct.

Why? To get to the tissues with the least expenditure of force.

How are arteries protected in the extremities? By being placed on the inside.

Where are they subject to pressure? At the joints.

What arteries diminish in size? Those that give off branches.

Exception? Abdominal aorta and vertebral arteries.

At what angle are arteries given off? At acute angles.

Exception? The posterior intercostal arteries.

At what angle are they given off? Right angle.

Where do arteries anastomose freely? In the brain.

Why? To protect the brain.

How are arteries divided? Large, medium, and small arteries.

Name the large arteries. All those larger than the common iliacs and carotids.

Properties of large arteries? Great strength and elasticity.

Name the medium sized arteries. All between the large and those one-twelfth to one-fifteenth inch in diameter.

Properties? Considerable strength and elasticity, and some contractibility.

Name the small arteries. All smaller than one-twelfth to one-fifteenth inch in diameter.

Properties? Little strength and elasticity, and great contractibility.

Coats of arteries? External, middle and internal.

Which coat is the same in all arteries? The internal.

Describe the external. White fibrous tissue.

Where does this coat cease? Just before the capillary system.

Function of outer coat? Strength.

Describe the internal coat. Simple homogeneous membrane.

Function of internal coat? To give a smooth surface for the blood to pass.

What is the middle coat of large arteries made up of? Yellow elastic tissue.

Are there any muscular fibres near the heart? No.

Why is the coat called " fenestrated membrane?" It has numerous foramina.

How does the middle coat change as you recede from the heart? Losses in elastic tissue and gains in muscular.

On which coat does the thickness of the artery depend? Middle coat.

What is the function of this coat? To give strength and elasticity.

Where is it wholly muscular? In the arterioles.

What are arterioles? The smallest arteries.

Describe their middle coat. A double layer of muscular fibre.

Which coat of the arteries is supplied by blood vessels? External.

What are they called? Vas vasorum.

Origin? From neighboring arterioles.

Nerves are distributed to which coat? Middle.

From what system do they come? Cerebro-spinal and sympathetic.

Name given to those from cerebro-spinal? Vaso motor dilators.

Effect on the artery if they are cut? It contracts.

Effect if stimulated? It dilates.

Effect if you cut the sympathetic? It dilates.

Effect if you stimulate the sympathetic? Contracts.

Does stimulation of the cerebro-spinal nerve to the artery cause it to dilate? It does not.

What does it counteract? The sympathetic.

What causes the artery to dilate? Pressure of the blood in the artery.

How is the blood supply to a gland during its functional activity? Very much increased.

Relation of the blood supply to the heart's action? Direct.

If the arteries were rigid tubes how would the heart have to pulsate during digestion? Very much faster.

What other force brings about the increased blood supply? Contractibility.

How does the blood flow from the heart? In an intermittent stream.

In the large arteries? Intermittent.

Medium sized and small arteries? Remittent.

Capillaries? Continuous.

What causes this change? Elasticity of the arteries.

Which coat? Middle.

What trouble would there be if blood were supplied to the tissues as it is sent out from the heart? Would have rupture of the capillaries and bad nutrition.

Give example of elasticity of arteries? They are distended at each systole.

Function of the elasticity? Aid in circulation, modify the force of the heart, and cause a continuous flow.

Does it assist in circulation? It does.

What conditions will cause more fluid to flow out of an elastic than rigid tube? The stream must be intermittent and there must be valves.

Why? More fluid gets into the tube.

Proof that arteries are elastic? Can distend them with water.

Function of the contractibility of arteries? Regulate the distribution of blood.

Proof of their contractibility? Puncture them and they will collapse.

What is "tonicity" of an artery? The power of keeping a certain amount of contraction.

When is it greatest? In winter.

What is the function of the arterioles? To regulate the distribution of blood.

What is the locomotion of arteries? The movements at each systole of heart.

Where is it seen? Where the arteries are tortuous and superficial.

What is the pulse? The force of the heart switched off in the arteries.

Where do you feel it? At the radial artery.

Why select this one? It is most superficial and rests on a bone.   •

How make the pulse greatest? Obstruct the blood beyond the point where you feel it.

When could the pulse not be felt? When the arteries are rigid from calcareous degeneration.

Name some different kinds of pulse? Soft, hard, wiry, thready, etc.

Time allotted to examine the pulse? One-seventieth of a minute.

Can the different characters be examined in this time? They can not.

How can you tell the characters? By means of the sphygmograph.

Describe the principle on which it is based? A lever which registers the pulse on a revolving cylinder.

What does the first ascension of the lever represent? The force of the heart.

Describe the descent? Slow and irregular.

What does the second rise represent? A second impulse.

Cause? Elastic recoil of the arteries.

What condition favors a dicrotic pulse? Low arterial pressure.

What is arterial pressure? The pressure of the blood in the arteries.

Time of the year it is greatest? Winter.

Why? The arterioles are more contracted than in summer.

How is arterial pressure away from the heart? Lowered.

Effect of hæmorrhage on arterial pressure? Lowers it.

Effect of asphyxiated condition? Increases it.

Why? The blood is dammed back into the arteries.

Effect of muscular contraction on arterial pressure? Increases it.

Why? By making pressure on the veins and compression of chest also aids arterial flow.

Name the depressor nerve of circulation. Nerve of Cyon and Ludwig.

Origin? Pneumogastric and superior laryngeal nerves.

Anastomoses? Sympathetic.

Distribution? To the heart.

Effect if cut? No effect.

Effect of stimulating the peripheral extremity? No effect.

Stimulating central end? Lowers arterial pressure.

Effect if cut great splanchnic? Lowers arterial pressure, but not so much.

What is the rapidity of flow of blood in large arteries? Twenty inches per second.

What does it represent? The force of the heart.

Condition after impulse? Eight inches per second.

What does this represent? Elastic recoil of the arteries.

Condition after the second impulse? Five inches per second.

Which diminish as you recede from the heart? Twenty and eight, intermittent flow.

Which increase? Five, continuous flow.

4

CHAPTER VIII.

CAPILLARIES.

Give the anatomical definition of capillaries? The smallest blood vessels.

Give the physiological definition? Those vessels that bring the blood close enough to the tissues to give off its nutriment.

Give their physiological anatomy? Have but a single coat, a homogeneous membrane.

Like what other coat? Like the endocardium.

Size of capillaries? One two-thousandth to one six-thousandth inch.

Where find smallest? Nervous and muscular tissue, retina of eye and Peyer's patches.

Where find those of medium size? Mucous membranes and mucous layer of skin.

Where largest? Bones and glands.

Where are they most abundant? Nervous and muscular tissue.

Give their general characters? They have no definite direction, uniform in diameter, anastomose freely and have but a single coat.

Describe the capillary circulation? Flows rapidly in the middle and has a still layer on the side.

Cause of the still layer? Capillary attraction.

What part of capillary do you find the red corpuscle? In the middle.

Where the white? In the "still layer."

What are stomata? Holes in the walls of the capillaries.

What is the capacity of the capillaries? Five hundred to eight hundred times that of arteries.

Cause of capillary circulation? Vis-a-tergo.

Rapidity of capillary circulation? One-fortieth to one-thirtieth inch per second.

Relation of rapidity to respiration? Direct.

Effect of cold? Contracts the capillaries and lessens the rapidity.

Effect of heat? Dilates the capillaries and increases the rapidity.

Effect of an irritant? Causes contraction at first.

Contraction is followed by what? Dilatation.

What is the first step of inflammation? Dilatation of the capillaries.

CHAPTER IX.

VEINS.

What is their only function? To carry the blood back to the heart.

How are they divided? Superficial and deep.

What are superficial veins? Those returning the blood from the surface.

What are deep veins? Those accompanying arteries.

Are all arteries accompanied by veins? They are not.

Which is not? Ext. carotid.

How many have the arteries of the extremities? Two.

Name those that have three? Brachial and spermatic.

What is the capacity of veins? Two and a half to four times that of the arteries.

Where greatest? In brain.

Does the amount of blood in the arteries vary? No.

Does it in veins? Yes.

Why do veins anastomose? For blood to be sure and get back to the heart.

Name the coats of veins. External, middle and internal.

Which are like those of the arteries? Internal and external.

Describe the middle coat. A double layer of muscular fibres.

Where have veins but one coat? In bones and the sinuses of the brain.

Which is thickest, artery or vein? Artery.

Why will a vein collapse when cut? On account of their thin walls.

What vein will not? The hepatic vein.

Why not? It is held open by adjacent hepatic tissue.

Which is strongest, artery or vein? Vein.

Exception? Splenic.

What is the outlet of the arteries? Capillaries.

Of veins? The heart.

Why should veins be stronger than arteries? Their outlet is smaller.

Veins are strongest in which extremity? Lower.

Proof of elasticity of veins? Distend them with water.

Proof of their contractility? Cut them and they contract.

Where do you find valves in veins? Everywhere, except in the cavities.

Where is the valve situated? Just beyond where a branch empties into the vein.

Describe the valve. Fold of the internal coat.

Where is the vein weakest? Between the valves.

Are there any valves in the spermatic vein? Yes.

Into what does the left spermatic vein empty? Left renal.

Into what does the left renal empty? Inferior vena cava.

At what angles do they both empty? Right angles.

What is varicocele? Dilatation of the spermatic veins.

On which side does it occur most? Left.

Why? The blood has two right angles to overcome and pressure of the sigmoid flexure on the left spermatic vein favors it.

What are piles? Dilatation of the hæmorrhoidal veins.

Causes that predispose to them? Looseness of the mucous membrane of the rectum and absence of valves in the veins.

What other outlet besides the liver have the hæmorrhoidal veins? Internal iliac vein.

Are the valves in the external jugular vein perfect? No.

What does pulsation in it indicate? Tricuspid regurgitation.

With what does the internal jugular vein unite? Subclavian vein.

What do they form? Vena innominata.

Why should there be perfect valves in the internal jugular vein? To protect the brain.

Why not necessary in the external? Reflex of blood does no harm.

The internal returns blood from what kind of tissue? Nervous tissue.

The external from what kind? Muscular tissue.

How is the flow of blood in arteries? Intermittent and remittent.

How is it in capillaries? Continuous.

How flow in veins? Continuous.

What is the course of the veins? Almost direct.

How are they arranged so that their onward flow shall not be impeded? They anastomose.

Under what circumstances would the flow be intermittent in veins? During the functional activity of a gland.

How is venous pressure compared to arterial? Inversely, in proportion.

Rapidity of flow in veins? Half as fast as arteries.

What is the main cause of venous circulation? Vis-a-tergo.

Name the causes that assist. Intermittent muscular contraction, gravity, inspiration and contraction of the veins.

When will muscular contraction aid it? When it is intermittent and the valves are perfect.

What act of respiration increases it? Inspiration.

How? It produces a suction force.

To what distance does this force act? To the liver below and the jugular veins above.

Why does it extend to the liver? Because the hepatic vein stands open.

What will a vein do if suction force is applied to it outside of the body? It will collapse.

Why not collapse when inside of body? The blood holds it open.

Give illustration of gravity assisting venous circulation? Hold hand up with veins distended and they empty.

Which coat of veins is contractile? The middle.

What is the function of valves? To prevent backward flow.

What conditions impede venous circulation? Continuous muscular contraction, gravity, expiration, contraction of the right side of heart.

What act of respiration is the thoracic cavity smallest? Expiration.

What kind of muscular contraction will retard? Continuous.

When is the blood not flowing into heart? During auricular systole.

What is the course of the blood then? Into the ventricles.

What kind of a box is the cranial cavity? An air-tight box.

Is it subject to atmospheric pressure ?   It is not.

Does the quantity of fluid in the cranial cavity vary ? No.

Does the quantity of blood vary ?   Yes.

What fluid equalizes the blood pressure ?   Cephalo Rachidian fluid.

Where do you find it ?   In the the sub-arachnoid space and central canal.

Where does the central canal communicate with the brain?   At the floor of the fourth ventricle.

When does this fluid rise up ?   During anæmia of the brain.

When does it recede ?   During hyperæmia of the brain.

During what act of respiration do the fontanelles bulge ? Expiration.

Why ?   Expiration aids arterial circulation and retards venous.

When recede ?   Inspiration.

Why ?   Inspiration aids venous and retards arterial flow.

Peculiarity of the vessels of the brain?   They are surrounded by large lymphatics.

Give the physiological anatomy of the corpus cavernosum?   It is surrounded by a delicate fibrous membrane which sends bands into the interior which divides it into cells.

Describe the arteries of the penis ?   Large and tortuous, have only one coat, and give off large arborescent branches.

What is the first thing in the process of erection?   Hyperæmia.

Describe it?   The vessels relax and the blood rushes in.

Is it a congestion ?   No.

Describe " derivative circulation?"   Where arteries communicate with the veins without the intervention of capillaries.

Where found ?   Elbow, knee, cheeks.

Peculiarities of pulmonary circulation? Artery is thinner and more distensible, capillaries large and allow the passage of any kind of blood, and the right side of the heart is weaker and valves are imperfect.

What is a circuit? Where blood goes from the heart to the tissues and back.

How many are there? Two.

Name them? Greater or systemic, and lesser or pulmonary.

Describe each? Greater begins at the left ventricle and ends at right auricle, the lesser begins at right ventricle and ends at left auricle.

In what time does the blood make both? Twenty-three seconds.

On what experiment is it based? Inject one jugular with ferrocyanide of potassium and watch for it in the one on the other side.

How much blood is sent out at each pulsation of the heart? Five ounces.

How many would it take for all the blood to pass through? Forty-five pulsations.

How long does it take for all to pass through? Twenty-three seconds.

What kind of increase of heart's action would increase the rapidity of the circulation? Physiological.

What kind of circulation do you have in fevers? Weak.

After death, where is the blood? In the veins.

Why? There is a post mortem contraction of the arteries.

CHAPTER X.

RESPIRATION.

Name the kinds of blood? Arterial, venous and capillary.

Which have been examined? Arterial and venous.

Give the characters of arterial? Bright red and will support nutrition.

Characters of venous blood? Blue black and will not support nutrition.

Character of capillary blood? Has not been examined.

From what do the principles come that are given off in the lungs? The blood.

What becomes of the loss in air in passing through the lungs? Goes to the blood.

What is the relation of oxygen to the circulation? It is direct.

Define respiration? The process by which the tissues receive oxygen and give off carbonic acid.

How does the oxygen reach the tissues? Through the blood.

Could tissue take on oxygen without blood? Yes.

Example? Detached portions of tissue take on oxygen.

What was the old idea as to where respiration took place? In the lungs.

Does the introduction of air into the lungs constitute respiration? No.

Example? Drain the system of blood and introduce air into the lungs, and will have symptoms of asphyxia.

What part do the lungs play? Receptacle for the air and blood.

In the lowest form of life how is respiration carried on? By the general surface.

How do fish receive oxygen? Through their gills.

What is the respiratory apparatus in higher animals? Lungs.

What conditions are necessary for respiration where there are lungs? Blood and air separated by a membrane through which the gases can pass.

On what is the rapidity of changes dependent? The rapidity of the circulation and activity of nutrition.

Name the openings in the pharynx? Mouth, posterior nares, Eustachian tube, larynx and œsophagus.

What kind of tube is the œsophagus? Flaccid.

What kind is the trachea? Rigid.

Why do they differ? Because they perform different functions.

What is the shape of the larynx? Hour-glass shape.

Describe the vocal cords? Fibrous bands that extend from the thyroid to the arytenoid cartilage.

On which is the mucous membrane thickest? On the superior.

What is the glottis? Opening from pharynx to larynx.

What is the epiglottis? Leaf-like lamella covering the glottis.

Is it essential? No.

Proof? Remove it and can do without it.

What is the rima glottis? The chink between the vocal cords.

When is it open? At inspiration and expiration.

What protects it during deglutition? The superior laryngeal nerve.

What is it a branch of? Pneumogastric.

What nerve animates the intrinsic muscles of the larynx? The inferior laryngeal.

What is it a branch of? Pneumogastric.

Effect if it is cut in young? Cartilages collapse and have suffocation.

What if cut in the adult? Borders on suffocation.

What is the nerve of voice? Recurrent laryngeal.

The power comes from what nerve? Spinal accessory nerve.

What root? Upper.

From what points does the trachea extend? From fifth cervical to third dorsal vertebra.

How many coats has it? Two.

Where are the rings? Between the two coats.

How many rings in the trachea? Fifteen to eighteen.

How far around do they extend? Two-thirds the way around.

What is at the posterior third? Trachealis muscle.

How many layers has the muscular coat of the œsophagus? Two.

Describe the external layer? It is in three fasiculæ, above which all blend in one below.

Describe the internal layer? It is thinner and circular in direction.

What kind of fibres in the upper part? Striated.

In the lower part? Nonstriated.

What kind of epithelium lines the œsophagus? Squamous.

What kind of glands has the mucous membrane? Small racemose.

What is the condition of the lower part of the œsophagus after the bolus has passed? It is constricted.

How many bronchial tubes are there? Two.

How do they divide? Dicotomously.

How terminate? In intercellular air passages.

Where? One-eighth of an inch from surface of lung.

Shape of rings in small tubes? Irregular.

How do the air passages terminate? In air cells.

What is the shape of an air cell? Oblong.

Where largest? Near the surface.

Give the parenchyma of the lungs? Air cells, intercellular air passages, connective tissue, blood vessels and lymphatics are held together by the subserous areolar tissue.

What is the external covering? Pleura.

What kind of membrane is it? Serous.

What do you find beneath it? Subserous areolar tissue.

Describe the elastic tissue? Small elastic fibres which have no definite direction but connect the air cells.

What give the lungs elasticity? Serous and subserous tissue and the elastic fibres.

Where do you find the richest capillary flexus? In the lungs around the air cells.

How arranged? To surround the air cells.

What kind of blood nourishes the lungs? Arterial.

What do the bronchial arteries supply? Bronchial tubes and lung substance.

Which lung has two arteries? Left.

What is bronchitis? Inflammation of mucous membrane of the bronchial tubes.

What is pneumonia? Inflammation of the parenchyma of lung.

Will bronchitis cause pneumonia? No.

Why not? They have different blood supply.

What acts make up respiration? Inspiration and expiration.

Which is active? Inspiration.

Why? It is produced by muscular contraction.

Which is passive? Expiration.

Peculiarities of the first rib? Shorter, more horizontal, and the sternal extremity is the most firmly fixed.

Direction of the other ribs? Oblique.

Which rib is longest? Eighth.

When the thorax is raised, in which direction is the thoracic cavity enlarged? In every direction.

What ribs approximate? Upper four.

Which separate? Lower eight.

Name the muscles of ordinary inspiration? Diaphragm, three scalenii, external intercostal, sternal portion of internal intercostals and levatores costarum.

Which is the most important one? Diaphragm.

Shape? Dome shaped.

How is it divided? Thoracic and lumbar portion.

Where is it muscular? At the edge.

How is it in the center? Tendinous.

Give the origin of the thoracic portion? Ensiform cartilage and six lower ribs.

Where is it inserted? Into the central tendon.

How is the lumbar part divided? Into two crura.

Which crus is largest? Right one.

Name the pillars in each one, also, their origin? Internal, from third lumbar vertebra ; middle, from second lumbar vertebra; external, from first, second and third lumbar vertebra ; intervertebral substance of the first, second, third and fourth, ligamentum arcuatus internum and externum and last rib.

Arrangement of the internal pillar? Figure of eight opening around the aortic and œsophageal openings.

Give the attachment of ligamentum arcuatum externum? From the apex of the transverse process of third lumbar vertebra to the last rib.

Arches over what muscle? Quadratus lumborum muscle.

Describe ligamentum arcuatum internum. Extends from apex to transverse process.

It arches over what muscle? Psoas magnus.

Describe the central tendon? It is the shape of a trefoil leaf, with the middle leaflet directed forward.

Which leaflet is largest? Anterior.

Name the openings in the diaphragm? Aortic, œsophageal and opening for the inferior vena cava.

Give boundaries of the aortic portion? Bone behind, internal pillars laterally and a tendinous arch in front.

What passes through it? Aorta and thoracic duct.

Why do they pass through there? So that no pressure is made on them by the contracting diaphragm.

Where is the œsophageal opening? In the muscular portion.

Why? To prevent regurgitation of food during inspiration.

Where is the opening for inferior vena cava? Tendinous portion between the right and middle leaflets.

Why? So no pressure is made on it.

What nerve supplies the diaphragm? Phrenic.

Give its origin? Third, fourth and fifth cervical nerves.

In inflammation of the diaphragm where is the pain? Shoulder, acromion process and clavicle.

Why? Nerves from the anterior cervical plexus supply both these parts.

What elevates and fixes the first rib? The scalenae muscles.

Direction of the fibres of the external intercostal muscles? Downward and forward.

What end of a rib is most firmly fixed? Vertebral end.

What is the origin of a muscle? A fixed point.

What is the origin of the external intercostals? Vertebral extremity of the ribs.

Why do they take their origin there? To take their origin from above.

How are the ribs pulled when they contract? Upward.

Name the ordinary auxiliary muscles of inspiration? Serratus posticus superior and sterno cleido mastoid.

Describe them? See anatomy.

Give the extraordinary auxiliaries? Levator angulæ scapulæ, trapezius, pectoralis major and minor, and serratus magnus.

When used? When respiration is greatly interfered with.

In what diseases is it used? Capillary bronchitis and asthma.

Describe these muscles? See anatomy.

What is expiration? Expulsion of air from the lungs.

What kind of an act? Passive.

What two forces produce it? Elasticity and muscular contraction.

Source of the elasticity? Lungs and thorax.

Prove that the lungs are elastic? Can distend them by blowing them up.

Why do the lungs go down in inspiration? There is a suction force produced when the diaphragm contracts.

Why do the lungs not fully collapse after death? The walls of the thorax prevent atmospheric pressure on them.

When are the thoracic walls most elastic? Early life.

Why? The ribs contain more animal matter and are consequently more elastic.

Why are fractures so frequent in old age? The bones are brittle in old age.

What kind of fractures do you find in young? Green stick fractures.

Why? The bones are very elastic.

Name the muscles of ordinary expiration? Osseous portion of the internal intercostals, infracostales, and triangularis sterni.

Describe them? See anatomy.

Ordinary auxiliaries? The abdominal muscles.

Describe them? See anatomy.

Name the types of respiration? Abdominal and superior and inferior intercostal types.

When use the abdominal type? In children.

What muscle performs it? The diaphragm.

What is the inferior intercostal type? Where the lower part of the thorax is most prominent in the movements.

When used? By the adult male.

What is the superior intercostal type? Where the upper part of the thorax is most prominent in the movements.

By whom is it used? By the adult female.

Why? So that during pregnancy the gravid uterus will not interfere with respiration.

What is the frequency of respiration? Average twenty per minute.

What is the relation of respiration to the heart's action? In the proportion of one respiration to every four beats of the heart.

Does this hold good in disease? Yes.

Exception? Pulmonary troubles.

Which breathes most rapid, male or female? Male.

Do the acts of respiration follow each other immediately? No.

Between which is there an interval? Expiration and inspiration.

Describe the inspiratory sound over the bronchial tube? Tubular in character, attains its maximum intensity quickly and ends abruptly.

Describe it over lungs substance? Vesicular in character, less intense, increases in intensity from beginning to the termination, and ends abruptly.

What causes the sound? The expansion of the air cells and the vibration of the air.

Describe the expiratory sound? Shorter, lower in pitch, soon attains its maximum intensity, and dies gradually.

Which act is the longest? Expiration.

Which sound is longest? Inspiratory sound.

Peculiarity of every fourth respiration? It is deeper than the rest.

At what period of life are the sounds most intense? Children.

What is it called? Puerile breathing.

Where do you find it in normal chest of male? Apex of right lung.

Why? The right bronchus is larger and enters the lung higher up.

Where do you hear it in normal chest of female? Apex of both lungs.

Why? Because female breathes mostly with the upper part of the lung.

What is the cause of snoring? Flapping of the uvula between the two currents of air from mouth and nose.

Cause of coughing and sneezing? Irritation of nasal or respiratory passages.

Give the mechanism? A deep inspiration followed by a convulsive expiration.

Give the mechanism of yawning and sighing? Deep inspiration followed by a rapid and audible expiration.

What is the cause of each? An effort to get more oxygen.

Give the mechanism of laughing and crying? Convulsive movements of the diaphragm, accompanied by contraction of the muscles of the face.

During what act of respiration do they occur? Expiration.

What is hiccough? A peculiar modification of inspiration.

Cause of hiccough? Dry food or effervescent drink.

Give the mechanism? Spasmatic contraction of the diaphragm, accompanied by spasmatic constriction of the glottis.

What is residual air? That that can not be expelled from lungs.

Amount? One hundred cubic inches.

5

Why is it necessary for air to remain in lungs? To make respiration continuous.

What is reserve air? That that can be expelled from lungs, but is not expelled at an ordinary expiration.

Amount? One hundred cubic inches.

Function? To be used in prolonged vocal efforts.

What is the tidal air? That used in ordinary respiration.

Amount? Twenty cubic inches.

What is complemental air? All that can be taken in over and above a normal inspiration.

Amount? One hundred and ten cubic inches.

What is the vital capacity? All that can be changed by a prolonged inspiration and forcible expiration.

It includes what airs? All except the residual.

Amount? Two hundred and thirty cubic inches.

Does air lose or gain in passing through lungs? Loses.

How much? One-seventieth to one-fiftieth.

Which element? Oxygen.

What causes tidal air to go down into the air cells? The diffusibility of gases.

Give the law of diffusibility of gases? They diffuse with inverse ratio to the square root of their densities.

What is the atomic weight of hydrogen? One.

Of oxygen? Sixteen.

What is the square root of each? Of one it is one ; of sixteen, it is four.

Which diffuses fastest? Hydrogen.

Is respiration continuous? Yes.

What was the old idea as to the function of respiration? To cool the blood.

Why? Because they found expired air warmer than inspired air.

What is the air? The atmosphere in which we live.

Is it a mixture or chemical compound? Mixture.

What is its composition? Nitrogen four parts and oxygen one part.

Effect of breathing dense air? Respiration is disturbed.

Effect of breathing light air? It is seriously interfered with.

What trouble would there be if the oxygen of the air were diminished from three to five per cent.? Death.

Effect of breathing pure oxygen? No effect.

What other gas will support respiration for a short time? Nitrous oxide gas.

Why was this gas called "laughing gas?" It produces delirium and exhilaration.

How will it cause death? It combines with oxygen in the corpuscle.

What gases produce death negatively? Hydrogen, nitrogen and carbonic acid gas.

Which kill positively? Arseniureted hydrogen, sulphuretted hydrogen and carbon monoxide gas.

How prove the loss of air in passing through lungs? Measure the oxygen inhaled and carbonic acid gas exhaled.

How many parts of oxygen in every one hundred parts of inspired air? Twenty-one parts.

How many uncombined in expired air? Sixteen parts.

Combined in expired air? Four parts.

What does it form? Carbonic acid gas.

How many parts lost in respiration? One part.

Amount of tidal air? Twenty cubic inches.

Number of respirations in a day? Twenty-six thousand.

Amount of air changed in a day? Three hundred cubic feet.

Amount of oxygen consumed in a day? Fifteen cubic feet.

What will increase the consumption of oxygen? Exercise, digestion and young age.

Diminish it? Fasting, sleep and old age.

What effect has mixing of hydrogen with it? Increases the consumption.

In what does the air gain in passing through the lungs? Carbonic acid gas, moisture, ammonia and organic matter.

How much carbonic acid gas in inspired air? Only a trace.

How much in expired air? Four per cent.

Name the elements of carbonic acid gas. Carbon and oxygen.

What is their origin? From the tissues.

How was the amount determined? Measuring it.

How much is exhaled in a day? Fourteen cubic feet.

How much carbon? Seven ounces.

What part of expiration is most expired? The last part.

Effect of rapid breathing on the amount of carbonic acid expired? It increases it.

State those things that will increase the exhalation of carbonic acid gas? Young age, males, digestion and exercise.

Those that will diminish it? Old age, females, fasting and sleeping.

Why less during the menstrual period? It is eliminated vicariously.

Give the relation between the inhalation of oxygen and exhalation of carbonic acid? They are in direct relation.

What becomes of the one part of oxygen left in the organism? It is thought to form water.

What favors the formation of water? Exercise.

What is the source of carbonic acid in expired air? From the tissues.

Name the forms in which it is found in the blood? Free and in the form of carbonates and bicarbonates.

How are these displaced? By the pneumic acid in the lungs.

How much water is exhaled in a day? One and one-fifth pound.

What is the origin of water? From the oxidation of hydrogen.

How prove ammonia is exhaled? Can smell it.

Proof of organic matter? Breathe into a sponge and leave it set, the organic matter becomes decomposed.

Does the air gain or lose in nitrogen? Gains a little.

What is hæmatosis? The changes that the blood undergoes in passing through the lungs.

What are the changes? Change of color, composition and properties.

What is the capital phenomena of respiration as regards the air in the lungs? The loss in oxygen and gain in carbonic acid.

The accessory phenomena? The gain in water, ammonia, nitrogen and organic matter.

What is the general difference between arterial and venous blood? In the amount of oxygen they contain. Arterial blood is uniform in composition and venous is not.

With what part of the blood is the oxygen in combination? The coloring matter in the red corpuscle.

What is the difference between the absorption of oxygen by water and the corpuscle? Corpuscle absorbs fifty times as much as water.

How is the absorption of oxygen by plasma? Twice as great as water.

How can you increase the carbonic acid absorbing power of the plasma? By adding phosphate of soda to it.

What was the old method of separating gases from the blood? By air pressure.

Fallacy of this method? The blood reabsorbed the gases.

Proof of post-mortem absorption of oxygen? Venous clot becomes red.

What is formed when oxygen is consumed? Carbonic acid gas.

What is the present method of separating gases from the blood? Add carbon monoxide gas to absorb the oxygen and then air pressure.

What is the difference between the amount of oxygen in venous and arterial blood? About twice as much in arterial.

Difference in the amount of carbonic acid? Venous has about four per cent. more.

Does the blood gain or lose in nitrogen in its passage through the lungs? Loses a little.

In what condition is oxygen in the blood? Unstaple combination with the hæmoglobin.

Why? So it can be readily given up to the tissues.

Condition of carbonic acid? In solution in the plasma.

Why? It is soluble in the plasma.

Give the mechanism of the interchange of gases in the lungs? Oxygen of the air passes down through the membrane into the blood and the carbonic acid is pushed out.

What is the relation of respiration to nutrition? Direct.

When respiration is disturbed, to what organ is it referred? To the lungs.

Does respiration take place there? No.

Does respiratory sense reside there? No.

Proof that it does not? Drain the system of blood and introduce air in the lungs, and will still have efforts at respiration.

Does it reside in tissue? No.

Where is it located? In the respiratory center.

Proof? Stimulate it and will cause efforts at respiration.

Cause of respiration? Want of oxygen in the respiratory center.

Is it direct or reflex? Direct.

What do you mean by direct? Originating in the center.

Why do the bowels act after death? Want of oxygen in the centers acts as an irritant.

How does respiratory sense differ from the sense of suffocation? Only in intensity.

Does the fœtus breathe? No.

Why not? Its blood is aerated in the placenta.

When would it breathe? When it does not get enough oxygen.

What are the functions of the placenta? It takes the place of the lungs and stomach.

What part of respiration takes place by the skin? One-fiftieth to one-fortieth.

Is it very important? No.

What is asphyxia? Want of oxygen by the tissues.

What are the symptoms? Full, strong pulse.

Why does pulse become full? To overcome the obstruction in the systemic capillaries.

At what period of life do we resist asphyxia longest? Childhood.

Why? Their demand for oxygen is less.

What is the most important question connected with asphyxia? The effect of carbonic acid on the system.

What was the old idea? That it killed positively.

Present idea? It kills negatively.

Reason for thinking so? Increase the oxygen in proportion to the carbonic acid and it will not kill.

On entering a crowded room, what sensation is experienced? Sense of oppression.

What becomes of it? It is lost.

Why? You become accustomed to it.

Trouble of rebreathing air? Danger of asphyxia.

What condition resists asphyxia longest? Debilitated condition.

Give illustration? Put a sparrow under a bell jar; in a short while put another under, the latter will die first.

Why will an invalid live longer than a healthy subject, both breathing poisoned air? The demand for oxygen is not so great in the invalid.

CHAPTER XI.

ALIMENTATION.

What becomes of all tissue? *Ans.* It wears out.

When organic tissue wears out, what does it produce? Effete matter.

What is the destination of inorganic matter? It is eliminated.

Trouble if the excrementitious products are not thrown off? Would have poisoning.

Can this wear be arrested? No.

What fluid furnishes elements for the formation of new material? Blood.

What does the blood contain? All the elements necessary for nutrition.

How do you keep it from becoming impoverished? By the introduction of food.

What is the want on the part of the system called? Hunger.

What is the first sensation of hunger? Appetite.

What kind of sensation is it? Not unpleasant.

What does it become if not appeased? Hunger.

Effect of introducing food? It is lost.

What do we have? A sense of satisfaction.

What if too much food is introduced? A sense of fullness.

How often do we get hungry? Three or four times a day.

On what does it depend? Habit.

What are the variations of hunger at different periods of life? Old and young get hungry most often.

Why? Is more tissue waste.

What time of the year get hungry oftenest? In winter.

Why? The food is more palatable and nutrition more active.

What effect has physical exercise? Increases it.

Why? It causes more tissue waste.

What condition is most favorable to appetite? To take a moderate amount of exercise.

What conditions diminish it? Fatigue and the administration of drugs like opium or alcohol.

What symptoms will you have if hunger is not satisfied? Pain in the epigastric region and frontal headache.

Effect of hunger over moral and intellectual faculties? It blunts them.

To what organ is hunger referred? Stomach.

Where located? In the tissues.

Proof? Introduce food into the stomach and let it run out at a fistula, will still have hunger.

Does the brain take cognizance of hunger? Yes.

Through what nerves? Sympathetic.

Prove that it does not through the pneumogastric. Cut the pneumogastric and will still have hunger.

How long can one live without food? Forty days.

What is thirst? Call on the part of the tissues for water.

What are the symptoms? Dryness of the throat and fauces.

What element is most necessary to life? Oxygen.

How long will one live without it? Three to five minutes.

Next most important to oxygen? Water.

How long can one live without water? Three to six days.

If thirst is not satisfied what symptoms would you have? Visceral inflammation.

Where is thirst located? In the tissue.

Proof? Thirst can be appeased without passing the water through the mouth.

What is aliment? Substances appropriated by the organism, or aiding in nutrition.

How divided? Direct and accessory.

Define direct? That is directly appropriated by the tissue.

Define accessory? That assists assimilation and retards disassimilation.

How is food divided? Organic and inorganic.

Name the division of the organic? Into nitrogenized and non-nitrogenized.

Name the nitrogenized food? Albumin, musculin, casein and fibrin.

Which is the most important one? Musculin or myosin.

What is musculin? Organic nitrogenized principle of muscle.

It is united with what? Inorganic matter.

How is it separated? By incineration.

Will flesh support life? Yes.

Why? It contains a variety of principles.

Will musculin support life? No.

Why not? Has no variety of principles.

Where is flesh digested? Stomach.

Where myosin digested? In small intestine.

How is albumin usually taken? In the form of eggs.

Will albumin support life? No.

Will eggs? Yes.

Why? They contain a variety of principles.

Where is albumin digested? Stomach.

What form of it is easiest digested? In whipped eggs.

How do we take casein in early life? In milk.

In adult life? Cheese.

Where is cheese digested? In the stomach.

What is the composition of this class of food? Carbon, hydrogen. oxygen. nitrogen, and a little sulphur.

Other name for this class? Albuminoids or protoids.

What is the origin of nitrogen in vegetables? From the saline matters.

What is the action of heat and moisture on the nitrogenized principles? Makes them undergo putrefaction.

Action of digestion on them? Changes them to peptones.

Action of acid on them? Changes them to peptones.

What is gluten? Nitrogenized proximate principle of cereals.

How is it obtained? Powder them and knead under a stream of water.

Color of gluten? Gray.

Why will it support life? It has a variety of principles.

What are they? Glutin, albumin, and fibrin.

Bread made from gluten is used in what trouble? Glycosuria.

Why? It contains no sugar-producing principle.

Name the non-nitrogenized class of food? Sugar, starch, and fats.

How do they differ from the nitrogenized? They have no nitrogen.

Which class is essential to life? The nitrogenized.

What is the supposed function of the non-nitrogenized? To produce heat.

Why think so? Because they are used most in winter.

Name the different kinds of sugar? Cane, milk, liver, honey, grape sugar.

What is their composition? Carbon, hydrogen and oxygen.

Properties of sugar? Sweet taste, soluble in water, burns with a caramel odor and leaves a carbonaceous mass.

Which is the sweetest sugar? Cane sugar.

What organ first produces it? Placenta.

In what blood do you find it? Blood from the liver to the lungs.

What is its destination? It is destroyed in the lungs.

Is it ever deposited? No.

When would it be eliminated? When a large amount is taken in.

Reasons for thinking sugar produces fat? They both contain the same elements, and people in sugar-growing regions grow fat.

Origin of starch? Mostly from vegetables.

Form usually found in? In the form of starch granules.

Describe a starch granule? It is depressed in the middle, which is called the hilum, and it has an investing membrane of nitrogenized material.

Test for starch? It gets blue with iodine.

Is it soluble in water? No.

Effect of boiling it with water? It swells up.

What is the first change of starch? Into dextrine.

Is it chemical? No.

Test for dextrine? Gets brown with iodine.

What is the next change? Into glucose.

Give the chemical change? Change into glucose.

Name the fluids that will digest starch? Saliva, pancreatic and intestinal juice.

Changes when subjected to them? Changed to glucose.

What is the origin of fats? Animal and vegetable kingdom.

Form in which it is introduced? Adipose tissue.

In what tissue is it not found? Bones, teeth and fibrous tissue.

What is the difference between butter and fats? Butter is softer and has no nitrogenized surrounding membrane.

Give the properties of fats? Oily feel and taste, neutral reaction, insoluble in water, soluble in ether and chloroform.

Action if subjected to high temperature and alkalies? They are decomposed into fatty acids and glycerine.

How are soaps formed? By the action of the fatty acids on the bases.

Where are fats digested? Small intestines.

By what fluid? Pancreatic juice.

By what are fats absorbed? Lacteals.

Are inorganic elements of food essential to life? Yes.

Name the most important one. Water.

How is it introduced? With all food and drinks.

Properties of pure water? Clear, colorless, odorless fluid.

What do you usually find in water? Organic matter and inorganic salts.

Symptoms one would have if not enough water? Thirst.

What if had too much? Dropsy.

Amount of water introduced per day? It is variable.

What influences the amount introduced? The kind of food and the amount eliminated.

Where do you find chloride of sodium? Everywhere except in enamel of the teeth.

How is it introduced? With food and drink.

How is it eliminated? Urine, sweat and mucus.

Function of chloride of sodium? Gives density to the plasma, aids osmosis and determines transudation.

Is there more in fluids or solids? Fluids.

Why are the bones brittle in old age? There is a deficiency in the elimination of the phosphates.

How is phosphate of lime introduced? With nitrogenized food.

Give its solubility? It is practically insoluble.

Found in fluids or solids? Solids.

In what bones do you find least? Those of sternum and the ribs.

Why? They must be elastic to aid in respiration.

Deficiency in phosphates gives what trouble? Rickets.

How is iron combined? With the hæmaglobine.

Proof of its exerting favorable influence over nutrition? Can soon notice its good effects when given an anæmic patient.

To what class does alcohol belong? Non - nitrogenized.

How is it eliminated? It is not eliminated normally.

What was the old idea? That it was eliminated by the lungs.

Why did they think so? They smelt it on the breath.

Present idea? It is destroyed in the organism.

When is it eliminated? When large amount is taken in.

What is the action of alcohol on the organism? First action is a stimulant action.

What is it followed by? Depression.

When is it not followed by depression? In disease.

Action on the elimination of urea? It lessens it.

On carbonic acid elimination? It lessens it.

How does alcohol act as a food? By lessening tissue waste.

Proof? There is less excrementitious products thrown off during its administration.

Effect of alcohol on the temperature? Lowers it.

How? It lessens oxidation of tissue by being itself oxidized, and it also increases the amount of blood to the heat losing area.

When do you give alcohol? Where you have a high temperature in long continued disease.

Why is it dangerous to get drunk in cold weather? The alcohol lessens the power of resistance and lowers the temperature.

Effect of coffee on the organism? Stimulant.

On what does its action depend? Caffein.

Action if taken in large quantities at night? Produces wakefulness.

What is its action on nutrition? It aids it.

What is the action of tea? Same as coffee.

On what does tea depend for its action? Theine.

How does tea differ from coffee? Tea has less of the active principle.

What is theobromine? Active principle of coco.

What is its action? Same as coffee.

How do the inferior animals take food? By instinct.

How does man take it in early life? By instinct.

What is the best guide as regards the kind of food? Appetite.

When do you quit eating? When the appetite is appeased.

Does this hold good in disease? No.

How are excretions influenced by food? Increased as the food is increased.

What excretion is discharged in the greatest quantity? Water.

Amount discharged in a day? Four and one-half pounds.

Functions of water? Acts as solvent for the salts, gives elasticity to cartilage, pliability to tendons, contractility to muscles and density to bone.

How can you make a theoretical diet? Introduce just enough to compensate the waste.

How much carbon is discharged in twenty-four hours? Eleven ounces.

How much nitrogen? Four and one-half ounces.

How much solid food introduced? Two or three pounds.

How is the amount of work influenced by food? Greatly.

Relation of the climate to the food introduced? Cold increases and heat retards it.

What trouble if confined to one article of food? Become emaciated.

Give illustration? In army or on board ship this happens.

How fatten animals? Give them a mixed diet.

What is the best diet for man? A mixed diet.

What is the treatment for scurvy? Give fresh vegetables.

What kind of muscular tissue is best suited to man? Beef.

Why? It has the greatest nutritive powers.

Does it differ chemically from other meat? No.

Does it differ physiologically? Yes.

Proof? Changing diet increases appetite.

Are soups, extracts, etc., nutritious? Very little.

Give the composition of roasted meats? Nitrogenized matter, fats, salines and water.

What kind of bread will support life? Brown bread.

Why? It contains a variety of principles.

What is the composition of bread? Gluten, nitrogenized matter, fat, salines and water.

Why are potatoes useful articles of diet? On account of the starch they contain.

What is their composition? Sugar, starch, fat, nitrogenized matter, salines and water.

Why is milk so useful? Has a great variety of principles.

What is the composition of butter? Olein and palmi-
tine.

Composition of milk? Casein, fat, sugar, inorganic matter and water.

Composition of eggs? Albumin, fatty and inorganic matter and water.

What kind of diet is necessary to support life? A mixed
diet.

CHAPTER XII.

MASTICATION.

What change do the inorganic principles undergo before introduction into the body? *Ans.* None.

Do they undergo change in the body? No.

What class of food needs preparation before introduction? Organic.

What is the digestive apparatus in the lowest form of
life? A simple pouch with a single opening.

Change as you ascend? It becomes more complicated.

In what animals is it most complicated? Herbivora.

Why? The food has so much waste matter.

What is the function of the colon? Receptacle for the
residue.

What is its length in carnivora? Three or four times the
length of the body.

In omnivora? Six or seven times the length of the body.

In herbivora? Ten or twelve times the length of the body.

What time is a full meal in digesting? Two to four hours.

On what does it depend? The kind of food and the fine-
ness of its comminution.

What is mastication? Chewing the food.

6

What is prehension? Carrying the food to the mouth.

In what animals are the organs of mastication most complicated? Herbivora.

Why? They take a large amount of bulky food.

In which is it simplest? Carnivora.

In what animals is digestion most prolonged? Carnivora.

How many teeth in each jaw? Sixteen.

Which teeth are best developed in carnivora? Canine.

Why? To tear the food.

Which in herbivora? Molars.

Why? To grind the food.

How is the tooth divided? Root, neck and crown.

What is the crown? That part above the gums.

What is the neck? Part between the crown and root.

What is the root? The part imbedded in the alveolar processes.

What parts make up a tooth? Enamel, dentine and cement.

Where do you find the enamel? Covering the crown.

Describe it? Hard, white, glistening substance.

What is the cuticle? A delicate membrane covering the enamel.

How is it demonstrated? Apply a mineral acid, it separates.

What does the dentine surround? The pulp cavity.

What surrounds the dentine? Enamel.

Describe the dentine? Is like bony tissue, but has no lacunæ nor canaliculi.

Where do you find the cement? Covering the root.

What is its other name? Crusta petrosa.

Describe it? It is true bony structure.

Where do you find the pulp cavity? On the interior of the tooth.

Give its anatomy? Blood vessels, nerves, lymphatics and connective tissue.

What causes a tooth to turn black? Decomposition of the pulp cavity.

Is the tooth then destroyed? No.

Give the perfected structures of a tooth? Enamel and dentine.

Why? Because they are never reproduced if destroyed.

How many incisor teeth? Four in each jaw.

What is their shape? Wedge.

How do they differ in the two jaws? In the 'upper jaw the middle teeth are largest, and in the lower, the lateral.

When do the incisors appear? Seventh or eighth year.

What is their function? To divide the food.

How many canine teeth? Two in each jaw.

What are the lower ones called? Stomach teeth.

The upper ones? Eye teeth.

Describe a canine tooth? Conical and pointed, and has a single long root.

When do they appear? Eleventh or twelfth year.

How many bicuspid teeth? Four in each jaw.

Describe them? Short, thick, and the crown is marked by two eminences.

When do they appear? Ninth or tenth year.

How many molars? Six in each jaw.

Names? First, second and third molar.

Describe them? Cube, rounded laterally and marked by three eminences on the crown.

When do they first appear? Sixth or seventh year.

The second? Twelfth or thirteenth year.

The third? Seventeenth to twenty-first year.

What is the last one called? Wisdom tooth.

Which jaw is broadest? Upper one.

Position of the upper teeth to the lower? They over-hang the lower.

Which jaw is movable? Lower.

Describe the lower jaw? Horse-shoe shaped, made up of a horizontal body and perpendicular ramus.

Describe the articulation? Condyle, is convex with the long diameter directed inward and backward.

What movements does it admit? In every direction.

Why is the condyle not dislocated when the jaw is brought forward? The interarticular cartilage prevents it.

Name the muscles of mastication? Buccinator, tempo-ral, masseter and two pterygoids.

Describe them? See Anatomy.

What muscles pull the jaw down? The elevators of the os hyoides.

Which elevate it? Temporal, masseter and the pter-ygoids.

Which bring it forward? Anterior fibres of the tempo-ral, oblique fibres of the masseter and the internal pterygoid.

Which bring it back? Posterior fibres of the temporal, digastric, mylo-hyoid and genio-hyoid.

Which pull it from side to side? External and internal pterygoids.

What is the nerve of mastication? Motor root of the fifth nerve.

What muscles pull the tongue forward? Posterior fibres of the genio-hyo-glossus.

Which pull it backward? Anterior fibres of the genio-hyo-glossus.

Which make it concave from side to side? The whole length of the genio-hyo-glossus.

Which make it convex from side to side? Hyo-glossus.

What is the function of the tongue in mastication? To keep the bolus between the teeth.

What other muscle assists? Orbicularis oris.

What nerve supplies both? Facial.

What is the function of the orbicularis oris? To close the mouth.

How do you know when the food is masticated? By the sense of touch.

What other digestive process takes place in the mouth? Insalivation.

What is saliva? Secretion of the salivary glands.

Name the salivary glands? Parotid, submaxillary and sublingual.

Where do you find the parotid gland? Behind and below the ear.

What covers it externally? Parotid fascia.

Internally? Stylo-maxillary ligament.

What bounds it posteriorly? Mastoid process of the temporal bone.

Anteriorly? Ramus of the lower jaw.

What class of glands does it belong to? Compound racemose.

What is its duct called? Duct of Steno.

Where does it empty? Opposite the second molar tooth.

Give the properties of parotid secretion? Clear thin fluid, coagulated by heat.

Function? To wet the bolus.

Old idea as to the cause of its secretion? Thought it was due to pressure of the muscles on the gland.

Present idea? Nervous influence.

How is the blood supply to the gland during the intervals of mastication? Just enough for nutrition.

During mastication? It is very much increased.

Why? To furnish watery element to wash out the active principle.

What is the inhibitory nerve of the blood vessels of the glands? Corda tympani.

Where do you find the submaxillary gland? In the submaxillary triangle of the neck.

Name the duct? Wharton's duct.

Where does it empty? On the side of the tongue.

Give the properties of the secretion? Clear fluid, gets gelatinous on cooling.

Function? Wet and coat the bolus.

When is the gland active? All the time.

Where is the sublingual gland? Under the tongue, one on each side.

Where does the duct empty? Side of the tongue near the frænum.

Describe the fluid? Clear, viscid, does not gelatinize on cooling.

Function? To coat the bolus.

Name the other glands that secrete saliva? Buccal, labial, lingual and pharyngeal.

What do you mean by saliva? Secretion of all the glands.

How much is secreted in a day? Two and one half to three pounds.

Give its properties? Colorless, viscid, alkaline frothy fluid, specific gravity 1004 to 1006.

Give its composition? Water epithelium, sulphocyanide of potash, inorganic salts, and organic matter.

On what does its action depend? Ptyalin.

How is it obtained? By precipitating it with pure alcohol.

It digests what kind of food? Starch.

Give the functions of saliva? Mechanical and chemical.

CHAPTER XIII.

DEGLUTITION.

Define it? *Ans.* Act of swallowing.

How many acts make it up? Three.

What parts are concerned? Mouth, pharynx and œsophagus.

Where does the pharynx begin? Base of skull.

Where does it end? Fifth cervical vertebra.

What separates it from the mouth? The soft palate.

How long is it? Four inches.

What shape is it? Funnel.

How attached above? Basilar process of the occipital bone and the first cervical vertebra.

With what is it continuous below? Œsophagus.

What is the shape of the anterior surface of soft palate? Concave.

Posterior surface? Convex.

To what is the upper border attached? To the hard palate.

Where do you find the free border? It is the inferior border.

How many pillars has the soft palate? Two.

Names? Anterior and posterior.

What is the direction of the anterior? Downward and forward, to the tongue.

Of the posterior? Downward and backward, to the pharynx.

What do you find between them? The tonsils.

What is the isthmus of the fauces? The space bounded by the base of the tongue below and soft palate above.

What are its lateral boundaries? The tonsils and pillars of the soft palate.

Where does the œsophagus begin? Fifth cervical vertebra.

Where end? Ninth dorsal.

What kind of a tube is it? Flaccid.

How long is it? Nine inches.

How many coats has it? Three.

Describe the external? Fibrous membrane.

What is the middle? Muscular fibres arranged in two layers.

Describe the longitudinal fibres? They are arranged in three fasciculi above and blend into one below.

What is their function? To shorten the tube.

Describe the circular? They are thinner and run transversely.

Function? Force the bolus down.

What kind of fibres are in upper part of œsophagus? Striated.

In lower part? Non-striated.

Describe the mucous membrane? Red and vascular above and pale below.

Which act of deglutition is voluntary? First act.

Do we usually take cognizance of it? No.

Give the mechanism of the first act? Mouth is closed, bolus pushed on the tongue, which carries it back.

Which part of the second act is voluntary? First half.

What is the second act? Passing through the pharynx.

Give the mechanism? The pharynx is elevated by the elevators of the os hyoides, and the constrictors force it down.

How is the second act influenced by cutting the nerve of mastication? It interferes with it, because you can't properly elevate the os hyoides.

Why can't you elevate the os hyoides? The mylo-hyoid and anterior belly of the digastric are paralyzed.

How influenced by cutting the facial nerve? Interferes with the elevation of the os hyoides.

Why? Stylo-hyoid and posterior belly of the digastric are paralyzed.

If cut the upper root of the spinal accessory? The second part of the second act is interfered with.

Why? The constrictors of the pharynx are paralyzed.

The pharyngeal muscles are supplied by what plexus? By the pharyngeal plexus.

What forms it? The pharyngeal branch of the pneumogastric and glosso-pharyngeal and the sympathetic nerves.

Where do the motor filaments come from? From the spinal accessory nerve.

How are the posterior nares protected during the second act of deglutition? By contraction of the posterior pillars of the soft palate and the superior constrictors of the pharynx.

How is the glottis protected? By the superior laryngeal nerve.

Name the reflex nerves of deglutition? Superior laryngeal and trifacial.

What becomes of respiration during the second act of deglutition? It stops.

What trouble if it does not cease? · Food would get into the air passages.

What is the third act of deglutition? Passing through the œsophagus.

Does it take place immediately? No.

Give the mechanism? The œsophagus is shortened by the longitudinal fibres, and then the circular slowly force the bolus down.

What is peculiar of the lower part of the œsophagus? It runs through the muscular portion of the diaphragm.

What is the function of the epiglottis? To protect the glottis.

Is it essential? No.

What condition is necessary for deglutition to take place? Must have something to swallow.

Is deglutition of air possible? Yes.

CHAPTER XIV.

DIGESTION.

What is it? *Ans.* Liquefaction of the food.

Where does it begin? Mouth.

What part takes place in the mouth? Mastication and insalivation.

What is the stomach? The most dilated part of the alimentary canal.

Where is it situated? Epigastric and left hypochondriac region.

What is the physiological definition for the stomach? A receptacle for the food and a place for digestion to take place.

How is the stomach divided? Surfaces, borders and extremities.

Name the surfaces? Anterior and posterior.

Name the borders? Superior and inferior.

What is another name for the borders? Greater and lesser curvature.

For the extremities? Cardiac and pyloric extremities.

What is the shape of the stomach? Bag-pipe.

Length? Fifteen inches.

Width? Five inches.

Capacity? About five pints.

What is in relation with the anterior surface? Diaphragm, left lobe of liver and abdominal walls.

What is in relation with the posterior? Diaphragm, transverse duodenum and colon, left kidney and suprarenal capsule and pancreas.

Name the coats of the stomach? Peritoneal, muscular and mucous.

Describe the peritoneal? Reflexion of peritoneum from the stomach.

Name the direction of the muscular fibres? Longitudinal, circular and oblique.

Where find the longitudinal? Along the lesser curvature.

Where find the circular? All over the stomach.

Where are they most numerous? At the pyloric extremity.

What do they form? Pylorus muscle.

Where find the oblique? Across the great extremity.

What is the thickness of the muscular coat? One-twenty-fifth inch.

What appearance has the mucous membrane when the stomach is empty? It is thrown in folds called rugæ.

What becomes of these rugæ during digestion? They become obliterated.

What appearance has the mucous membrane when washed? It is marked by numerous pits.

Where is the mucous membrane thickest? At the pyloric extremity.

What kind of epithelium lines it? Columnar.

What glands do you find in it? Acid and peptic glands.

Where do you find the peptic glands? Near the cardiac opening.

Where find the acid glands? Throughout the stomach.

The secretion in what part of the stomach is acid? Greater pouch.

What lines the peptic glands? Peptic cells.

What lines the secreting part of the acid glands? Peptic cells.

What lines the acid part? Acid cells.

What is the function of acid glands? To secrete hydrochloric acid.

When is pepsin produced? All the time.

When is acid produced? When food is in the stomach.

What do peptic cells produce? Zymogen.

What changes it into pepsin? The hydrochloric acid.

What lines the excreting part of the gland? Columnar epithelium.

What is the color of a peptic cell? Light pink.

Of the acid cell? Dark.

In what period of life do you find closed follicles? In children.

Where are they most abundant? In the great curvature.

What reaction is the stomach during the intervals of digestion? Alkaline.

During digestion, how is it? Acid.

What is gastric juice? The secretion of the glands of the stomach.

When is it discharged? During digestion.

How is it obtained? By making a fistula.

Name three ways in which you could obtain a digestive fluid? Mixing acid and pepsin, making a gastric fistula, or making an infusion of the mucous membrane of the stomach.

What part of the mucous membrane would you select to make an infusion? The pyloric extremity.

Why? The peptic glands are found there.

What two elements are used to make it artificially? Acid and pepsin.

How can you get pure gastric juice? By making a fistula.

What is the best kind of food to introduce? Savory food.

How should it be introduced? By the mouth.

Why? It mixes with saliva which aids the flow of gastric juice, and passing through the natural channel also aids it.

Why not use a local stimulant? If articles are given which the animal relishes it aids the flow.

What is the best stimulant? Natural food.

What influence has saliva? It increases it.

When does the secretion of gastric juice cease? When the food is of a pultaceous mass.

How much gastric juice secreted in a day? Six to fourteen pounds.

What becomes of it? That not used in digestion is re-absorbed.

What part do the inorganic salts play in digestion? Very little.

Give the properties of gastric juice? Clear amber fluid, non-viscid, acid in reaction, specific gravity 1005 to 1009.

Give its composition? Water, acid, pepsin, chlorides and phosphates.

What is its active principle? Pepsin.

When is it produced? All the time.

By what? Epithelium of the glands.

Is it a secretion? Yes.

Why? It is manufactured in a gland.

To what class of secretions does it belong? Transitory.

What is the object of that class? To aid in digestion.

How do you obtain pepsin? Make an acidulated infusion of the mucous membrane.

Give its properties? Grayish translucent scales of characteristic odor.

In what is it soluble? Water and weak alcoholic solutions.

What kind of acid has gastric juice? Hydrochloric.

What part of the gastric juice is necessary for digestion? Acid and pepsin.

Could there be too much pepsin ?   No.
Could there be too little ?   Yes.
Could there be too much acid ?   Yes.
What would you do for it ?   Give an alkali.
When would you give it ?   After meals.
What action would it have if given before meals ?   It would increase the acidity.
If there is too little acid, what do ?   Give them an acid after meals.
Why give it after the meal ?   To make up for the deficiency.
Why think you need no special acid?   Get the same effect from any of them.
In what way does chloride of sodium aid digestion ?   It aids the flow of gastric juice.
What class of food is digested in the stomach ?   Nitrogenized.
Which is the most important one ?   Musculine.
Where is it digested ?   Small gut.
What is the action of gastric juice on meat ?   It digests the fibrous tissue.
Where is myosin digested ?   Small gut.
Where is albumin digested ?   Stomach.
What kind of food is easiest digested ?   Whipped eggs.
What is the action of gastric juice on fibrin casein and nitrogenized vegetables ?   It digests them.
Action on cane sugar ?   Digests it.
What fluids digest cane sugar?   Gastric and pancreatic juice.
What part of the gastric juice ?   The acid.
What are peptones ?   Products of stomachic digestion.
Different kinds ?   Acid albumin, or albuminate, and propeptone, or hemialbumose, and true peptones.
How do they differ from albumin ?   They are osmatic and not coagulated by heat as albumin is.

If albumin is injected into the veins, what becomes of it? It is rejected by the kidneys.

What if cane sugar is injected into the veins? Rejected by the kidneys.

What becomes of albuminose if injected into veins? It is deposited.

How many fluids digest nitrogenized principles? Two.

Name them? Gastric and pancreatic juice.

Mostly digested by which one? Gastric juice.

What is the action of gastric juice on starch? It digests the nitrogenized membrane.

What fluids digest starch? Saliva, pancreatic and intestinal fluids.

The greater part is digested by which one? Saliva.

What is the action of gastric juice on fats? Digests the nitrogenized membrane.

Where are fats digested? Small gut.

What is the time of stomachic digestion? Two to four hours.

On what does it depend? On the kind and amount of food.

What food is easiest digested? Eggs, milk, fish, tripe, etc.

What condition is most favorable for digestion? A moderate amount of exercise.

What retards it? Fatigue.

Name the motor nerve of the walls of the stomach? Pneumogastric.

Proof? Cut it and have paralysis of the stomach.

Name the inhibitory nerve to the blood vessels? Pneumogastric.

If the pneumogastric be cut, how is the blood supply? Lessened.

If stimulated? Increased.

Why will cutting it stop digestion? It cuts off the blood and paralyzes the walls of the stomach.

What is the effect of cutting the sympathetic? It increases the blood supply.

If stimulate the sympathetic? Lessens the blood supply.

What change is there in the position of the stomach when it is distended? The anterior surface becomes superior.

Why does it cause dyspnœa? Due to pressure on diaphragm.

Where does the bolus enter the stomach? Cardiac extremity.

Course? Along the greater curvature and then back along the lesser.

Time it takes? Three minutes.

Why are the walls of the stomach paralyzed when the pneumogastric nerve is cut? It is the motor nerve of the stomach.

What is its origin? Between the corporas olivaria and restiformia.

In what animals is regurgitation of food common? Herbivora.

Is it physiological in man? No.

Give the mechanism? Relaxation of the cardiac extremity followed by contraction of the stomach.

What is emesis? Vomiting.

How are emetics divided? Centric and local.

What is a local emetic? One that acts by its presence in the stomach.

What is a centric one? It acts through the nervous system by an action on the nerve centers.

What is the first thing in emesis? Deep inspiration.

What is next? Fixation of the diaphragm.

How is the diaphragm fixed? By closure of the glottis.

What part of the stomach relaxes? The cardiac extremity.

What part contracts? The rest of it.

What is the last act in emesis? Expulsion.

CHAPTER XV.

SMALL INTESTINES.

What is the most important part of the digestive system? *Ans.* The small intestines.

Why is it so called? Because there is a larger one.

Where does the small gut begin? At the stomach.

Where end? Right iliac fossa.

How long is it? About twenty feet.

How is it divided? Duodenum, jejunum and ileum.

Why was the duodenum so called? It is the length of the breadth of twelve fingers.

How long is it? Nine inches.

What shape is it? Horse-shoe.

How is its concavity directed? To the left.

What is found in it? The head of the pancreas.

How is the duodenum divided? Into ascending, descending and transverse portion.

What is the direction of the first part? Upward and backward.

The relation of peritoneum to the first part? Completely surrounds it.

Give the relation of the first part? Above is the liver and neck of the gall-bladder, in front is the great omentum

7

and abdominal walls, behind is the lesser omentum and its vessels.

What is the direction of the second part? Down to the fourth lumbar vertebra.

Relation to the peritoneum? It is behind the peritoneum.

Give the relation of the second part? In front is the colon, behind, the inferior vena cava and kidney and ductus communis choledochus, on the right the ascending colon, on the left the head of the pancreas.

What duct empties into it? Ductus communis choledochus.

Where? At the posterior and internal aspect.

Where does the third part begin? Upper border of the fourth lumbar vertebra on the right side.

What is its course? Transversely across the third lumbar.

Where does it end? Left side of the second lumbar vertebra.

How long is it? Half the length of the duodenum.

Relation of the peritoneum? In front and on the sides.

What part of the duodenum is most firmly fixed? Third part.

Which is most movable? First part.

What is the jejunum? The second part of the small gut.

Why was it so called? It was found empty after death.

What is its length? A little less than two-fifths of the entire length.

Where does it begin? Left side of the second lumbar vertebra.

Where end? At the ileum.

What does ileum mean? Twisted.

How long is the ileum? A little more than two-fifths of the entire length.

Name the coats of the small gut? Serous, muscular and mucous.

Describe the external? A serous membrane that binds the guts to the vertebral column.

What is the mesentery? The serous membrane that binds the guts to the spinal column.

Where does it begin? At the second lumbar vertebra.

Where end? Right iliac fossa.

How many borders has it? Two.

Name them? Free and attached.

Length of each? Free is eighteen feet and attached is six inches.

What is the function of the mesentery? To bind the guts to the spinal column and to transmit the blood vessels and lymphatics.

What is the middle coat of the guts? Muscular.

What is the function of the longitudinal fibres? To shorten the gut.

Of the circular? To constrict and force the bolus down.

Where are they most numerous? In the upper part.

What is the difference in the mucous coat of the stomach and small gut? It is thinner in the gut.

Where is it thickest? In the duodenum.

Name the glandulas elements of the small gut? Valvulæ conniventes, glands of Brunner, follicles of Lieberkuhn, single and agminated glands and intestinal villi.

Where do you find Brunner's glands? In the upper half of the duodenum.

What kind of glands are they? Compound racemose.

As regard the coats of the gut, where do you find them? In the sub-mucous coat.

What size are they? One-tenth inch.

Give their physiological anatomy? A large number of little follicles all centering into one duct.

What kind of fluid do they secrete? Clear, viscid, alkaline fluid.

What is its function? It is not known.

Where do you find the valvulæ conniventes? From the middle of the duodenum to the lower third of the ileum.

How far around do they extend? One-half to one-third the way around.

What do you find between them? Connective tissue and blood vessels.

What is their function? To increase the surface exposed.

How much does it increase it in the jejunum? It doubles it.

How much in the ileum? One-sixth.

Where do you find the follicles of Leiberkuhn? Throughout the large and small gut.

What shape are they? Tubular.

What part is not occupied by these glands? Parts occupied by other glands.

Give their anatomy? They are simple inversions of the mucous membrane.

How do they terminate? In a single or double blind extremity.

How long are they? One seventy-fifth inch.

What is their diameter? One three hundred and sixtieth inch.

Function? To secrete intestinal juice.

Where do you find the villi? Throughout the small gut.

They have chiefly to do with what? Absorption.

On which side of the ileo-cœcal valve do you find them? On the iliac side.

What appearance do they give the mucous membrane? Velvety.

Where are the villi most numerous? In the duodenum.

How many to the square inch in the small gut? 7000.

What shape are they in the human? Flattened cones.

Describe them in the duodenum. Flat and short.

In the jejunum. They are longer.

In the ileum. Are shorter.

What form predominates in the upper part of the gut? The short, flat variety.

Where are they longest? In the jejunum.

How long are they? One-thirtieth to one-twentieth inch.

What is their diameter at the base? One-seventieth to one hundred and twentieth inch.

How are they formed? By an eversion of the mucous membrane.

Describe the membrane? A structureless basement membrane covered by epithelium.

Where do you find the epithelium? On the outside.

What kind? Columnar.

Where do the arteries enter the villus? At the base.

How many? Six to twelve.

How do they terminate? In capillary loops.

Describe the veins? At first there are four or five, but they all terminate in one.

Direction of the muscular fibres of the villi? Longitudinal.

Describe them? They run half way between the center and periphery of the gland.

What is their function? To shorten the villi.

How do lymphatics begin? In lymph spaces.

What are they called? Lacteals.

Why? They carry a milk-like fluid.

What fluid is it? Emulsified fats.

Where find the solitary glands? Throughout the large and small gut.

Where do you find the agminated glands? In the ileum.

What is their other name? Peyer's patches.

Where are they found? Mostly in the ileum.

On what part of the gut are they? Opposite the attach-ment of the mesentery.

Are they always present? They are.

What shape are they? Oblong.

Length? Three-fourths to one and one-half inch.

Width? One-half to three-fourths inch.

Where are they largest? In the lower part of the ileum.

How many are there? Eighteen to twenty.

Where may they be found? In the jejunum and ileum.

How many varieties are there? Two.

Name them? Smooth and prominent.

Describe the more prominent variety? They are covered by mucous membrane which is thrown in folds.

What is the arrangement of the valvulæ conniventes near their border? They stop.

Why are Peyer's patches not always found? The smooth variety are overlooked.

What kind do you have in feeble persons? Smooth variety.

What appearance have the villi on the large kind? They are large and prominent.

What is the shape of the follicles? Pear-shaped.

How is the pointed extremity directed? Toward the in-side of the gut.

What is the peculiarity of the mucous membrane just above the follicle? Has a small opening in it.

Do you find this in the small ones? No.

What is the diameter of the follicles? One seventy-fifth to one-twelfth inch.

Describe the surrounding membrane? Simple homoge-neous membrane.

What does it contain? Semi-fluid grayish contents, cells, blood-vessels and lymphatics.

What is the arrangement of the blood-vessels? A network around each follicle.

What part of their anatomy has not been demonstrated? Lymphatics.

Why is it thought that they contain lymphatics? The absorption of fats can be accounted for in no other way.

What is the function of Peyer's patches? Has to do with absorption.

In what trouble are they ulcerated? Typhoid fever.

Why are the mesenteric glands enlarged? Because the lymphatics coming from Peyer's patches pass through them.

Name the last digestive fluid? Intestinal juice.

What produces it? The glands of the small intestine.

How is it obtained? Empty a part of the gut and ligate each end and allow the fluid to collect.

Give its properties? Viscid fluid of rose tint and alkaline reaction.

How much solid constituents has it? Five to five and one-half per cent.

The digestion of what food is completed by it? Nitrogenized food and starch.

CHAPTER XVI.

PANCREAS.

What does pancreas mean? *Ans.* All flesh.

Where is it found? In the abdominal cavity, opposite the second lumbar vertebra.

What other names has it? Sweet bread or abdominal salivary gland.

What shape is it? Wedge.

How is it divided? Head, body and tail.

How long is it? Seven inches.

Weight? Four or five ounces.

Where do you find the head? In the concavity of the duodenum.

Where find the tail? In apposition with the spleen.

Name the surfaces? Anterior and posterior.

Borders? Superior and inferior.

What grooves the upper border? The splenic vessels.

What grooves the lower border? Superior mesenteric artery.

What kind of a gland is the pancreas? Compound racemose.

Where does the duct begin? At the tail.

How many at the beginning? Two.

Where do they unite? At the junction of the large third with the middle.

Nearest which surface is the duct? Anterior.

Destination of the duct? Empties into the ductus communis choledochus.

When is the gland active? All the time.

How can you obtain the fluid? Make a fistula in the duct.

How much is secreted in an hour? One hundred grains.

Give its properties? Viscid, slightly opaline fluid, specific gravity 1040.

Give its composition? Water, organic matter, chlorides, phosphates and carbonates.

Name the active principles? Amylopsin, tripsin and steapsin.

What does each digest? Tripsin digests the nitrogenized food, steapsin the fats and amylopsin the starch and cane sugar.

How do these substances differ from albumin? Their dried alcoholic precipitate is soluble in water, while that of albumin is not.

What is the action of pancreatic juice on starch? Changes it to sugar.

What is the action on cane sugar? Changes it to glucose.

On nitrogenized food? Digests it.

On fats? Emulsifies them.

Name the only fluid that will form a complete emulsion? Pancreatic juice.

What is an emulsion? Fine particles of fat held in suspension by some gummy substance.

By what is it absorbed? Lacteals.

What kind of action takes place in the digestion of fats? Mechanical.

What is the action of pancreatic juice on fats outside of the organism? Decomposes them into fatty acid and glycerine.

What would you suspect if called to see a patient with fatty diarrhœa? That there was some trouble with the pancreatic juice.

CHAPTER XVII.

BILE.

What is bile? *Ans.* A secretion and excretion of the liver.

What part is a secretion? Biliary salts and sugar.

Which is an excretion? Cholesterin.

Why is it discharged continuously? To get rid of cholesterius.

Is bile essential to life? It is.

How long will a dog live when not discharged into the small gut? Forty days.

With what symptoms would he die? Symptoms of inanition.

How many functions has bile? Three.

Name them? Nature's purge and disinfectant and to aid in digestion.

What is jaundice? A deposit of bile pigment in the tissues.

Name the different kinds? Hæmatogenous and hepatogenous.

Cause of hæmatogenous variety? Impoverished or poisoned condition of the blood.

What would be the treatment for it? To improve the condition of the blood.

What is the cause of hepatogenous variety? Where it is due to obstruction of the bile duct.

What would be the treatment? To remove the obstruction.

What is peristalsis? A worm-like movement of the gut from above downward.

What is anti-peristalsis? Worm-like movement from below upward.

What excites the regular movement of the guts? The presence of bile.

Where is the contraction most vigorous? In the upper part.

Why? There is more bile there.

What is the function of the gases of the small gut? To keep the walls distended.

What causes the bowels to act after death? Want of oxygen in the centers acts as an irritant.

What is the motor nerve of the small gut? Sympathetic.

Which is the inhibitory nerve? Pneumogastric.

How are the movements influenced by cutting the pneumogastric nerve? They are not affected.

How is the blood supply influenced? Lessened.

What are the symptoms if the great splanchnic is cut? Loss of motion of the guts.

How is the blood supply influenced? It is increased.

CHAPTER XVIII.

LARGE INTESTINES.

How long are they? *Ans.* Four to six feet.

Where do they begin? Right iliac fossa.

Where end? At the anus.

What is their position to the small guts? Surround the small gut.

What is the largest part of the large intestine? Cæcum.

What are some other names for it? Caput coli or blind pouch.

Where do you find it? In the right iliac fossa.

Size? Two and one-half to three and one-half inches long.

What empties into it? The ileum.

What is the relation of the peritoneum to it? It is in front and on the sides.

What is typhlitis? Inflammation of the cæcum.

What is perityphlitis? Inflammation of the peritoneum around the cæcum.

Where do you find the right colon? On the right side, extending from the cæcum to the liver.

What other name has it? Ascending colon.

Relation of peritoneum to it? Is in front and on the sides.

Where does the right colon end? Under the liver.

Name the flexure? Hepatic flexure.

What is the arch of the colon? The transverse colon.

Relation of peritoneum? Surrounds it.

Where do you find the splenic flexure? In opposition with the spleen.

Relation of peritoneum to left colon? In front and on the sides.

Where does the left colon end? At the crest of the ileum.

What is the shape of the sigmoid flexure? Like the italic letter *S*.

Relation of the peritoneum to it? Surrounds it.

What is the smallest part of the large gut? The opening of the sigmoid flexure into the rectum.

Where does the fæces accumulate? In the sigmoid flexure.

What is the length of the rectum? Nine inches.

Why is it called rectum? It is comparatively straight.

Where does it begin? At the left sacro-iliac symphysis.

Where does it end? A half inch in front of the coccyx.

How is the rectum divided? First, second and third portion.

What is the length of the first part? Half the length of the entire rectum.

Relation of the peritoneum to this part? Surrounds it.

Give the relations of the first part? Behind pyriformis muscle and origin of the sacral plexus of nerves, in front by the bladder in the male and uterus and its appendages in the female.

Describe the second part? Runs downward in the concavity of the sacrum.

Relation of peritoneum to it? Covered in front on the upper part.

Relation of the second part? In front by male, base of the bladder vesicula seminales, vas deferens, in the female the posterior wall of the vagina.

Describe the third part? Runs backward and terminates a half inch from the coccyx in the anus.

Where do you find the ilio-cæcal valve? Between ileum and cæcum.

How is it formed? By folds of the walls of the gut.

What is the direction of the slit? Horizontal.

How many lips has it? Two.

How is their concave surface directed? Toward the ileum.

What is the effect of pressure on the cæcal side? Closes the valve.

On the iliac side? It opens it.

Give the anatomy of the valve? Fibrous tissue and circular muscular fibres, all covered by mucous membrane.

What keeps it from pulling out? It is continuous with the muscular coat of the gut.

What is the function of the valve? To keep the fæces out of the small gut.

Name the coats of the large gut? Peritoneal, muscular and mucous.

Where is the peritoneal coat complete? Transverse colon and sigmoid flexure.

What is the arrangement of the muscular fibres? Circular and longitudinal.

Which layer is external? The longitudinal.

In how many bands is it arranged? Three.

Position in the ascending colon? Anterior and two latter posterior.

What is their position in the transverse colon? The anterior band becomes inferior and the others posterio, superior and inferior.

How are they in the descending colon? Same as in the ascending.

How, in the rectum? Two become blended into one, so it leaves two bands.

How, in the last part of the rectum? They all blend into one and surround the rectum.

Why does the large gut appear puckered? The longitudinal fibres are shorter than the gut.

Describe the circular fibres in the cæcum and colon? Form only a thin layer.

What is their arrrangement in the rectum? Form the sphincter muscle.

How does the mucous membrane of the large gut differ from that of the small? Paler, thicker, firmer and more closely attached.

Name the glands of the large intestine? Follicular and utricular.

Describe the mucous coat in the last part of the rectum? It is loose and thrown into folds.

Describe the veins at that part? They are very numerous.

What is dilatation of them called? Piles or hæmorrhoids.

Have they any valves? No.

What other outlet besides the liver has the veins of the guts? The internal iliac vein.

Which hæmorrhoidal unites with it? The middle one.

Give the characters of the fluid secreted by the glands of the large gut. A glairy mucus.

Has it any influence over digestion? It has not.

Where are the contents of the intestines most fluid? Small gut (duodenum).

Where most consistent? Sigmoid flexure.

What is its color in the jejunum? A bright yellow.

What color in the sigmoid flexure? Dark brown.

What causes it to change? Absorption of the watery material.

What is the contents of the small guts called? Chyme.

To what is fermentation in the intestinal canal due? To the action of micro-organisms.

How many micro-organisms has Pasteur isolated? Seventeen different kinds in the mouth.

What is the action on albumin? Dissolves it.

What kind of fermentation do you have in the guts? Putrefactive.

Name some substance resulting from intestinal fermentation? Indol, skatol, phenol, etc.

Name some drugs that will arrest the action of indol, etc.? Calomel, salicylic acid.

What is the contents of the large gut called? Fæces.

How does it differ from that of the small? In odor and color.

On what does the color depend? Bile.

How much fæces eliminated in a day? Four and one-half ounces.

On what does the consistence depend? On the length of time it has been in the gut.

Name some of the micro-organisms of the large gut? Bacterium coli commune, bacterium lactis aerogenes, the large bacilli of Bienstock, etc.

What is excretin? A crystalline substance extracted from the fæces.

What is stercorine? Modified cholesterin.

When is cholesterin discharged from the rectum ?. During fasting.

Putrefactive process is most marked in what gut? Large gut.

What gives the fæcal odor ? The products of intestinal putrefaction.

How can this be produced out of the organism ? By the action of pancreatic juice on albuminoids.

What is the action of pancreatic juice in an alkali medium on tripsine peptones ? Changes it into leucin, tyrosin, hypoxanthine and asparaginic acid.

What if the action carried further ? Forms indol, skatol, and phenol.

How does the contraction of the large gut differ from that of the small ? It is less vigorous and rapid.

What is the function of the longitudinal fibres ? To shorten the gut.

What causes matter to pass through the cæcum and ascending colon ? The pressure from the small gut.

CHAPTER XIX.

DEFÆCATION.

Define it ? *Ans.* Expulsion of fæcal matter from the rectum.

How often does it occur ? Once a day.

What is the best time to defæcate ? In the morning.

Why ? On arising the blood supply to the guts is increased and also the movements of the gut.

Where does the fæcal matter accumulate ? In the sigmoid flexure.

What causes the desire to evacuate? Pressure on the nerves in passing into the rectum.

How can you prove that fæces does not accumulate in the rectum? The rectum is found empty.

What is the narrowest part of the large gut? Opening from sigmoid flexture to rectum.

What is it called? Sphincter O'Beirne.

What do you mean by a figured evacuation? Where the fæces is moulded to the shape of the gut.

What is tenesmus? Frequent and painful desire to evacuate.

What is tormina? Colicky pains.

Describe the first part of evacuation? Fæcal matter is forced through the sphincter O'Beirne.

What resists the matter? The sphincter muscle.

How is it dilated? By the pressure of the fæces.

What is the function of the levator ani? To assist in dilatation of the sphincter.

What other muscles help perform the act? The abdominal muscles.

Give the mechanism? Anus is dilated, rectum becomes almost straight and the fæces is forced down.

What interrupts the discharge? Contraction of the sphincter ani.

Is the internal sphincter voluntary? It is not.

Name the gases of the stomach? Oxygen, hydrogen, nitrogen and carbonic acid gas.

Name the gases of the small intestine? Hydrogen, nitrogen and carbonic acid gas.

In large gut? Hydrogen, nitrogen, carbonic acid and carburretted hydrogen.

What is their origin? They are given off during the fermentation of the food.

What is their function? To keep the guts distended.

8

CHAPTER XX.

ABSORPTION.

What is digestion? *Ans.* Liquefaction of the food.

Where does it begin? In the mouth.

Where does it end? In the small intestines.

What is the first object of digestion? To liquefy the food.

What is the second object? To prepare it for absorption.

What is the first transformation? Starch into sugar.

What is absorption? Getting into the blood and lymph.

Where does it begin? In the mouth.

Where end? In the rectum.

Proof of its taking place in the mouth? The chewing of tobacco makes some persons sick.

Absorption is carried on by what vessels? Veins and lymphatics.

What does the blood absorb? All that will form a homogeneous mass with the blood.

What is absorbed by the lymphatics? All the rest.

Proof of absorption in the rectum? Get physiological effects from an enema.

Where does the greater part take place? In the small gut.

What change does the alimentary mass undergo in passing down the alimentary canal? It becomes more consistent and the biliary salts disappear.

Why are lymphatics hard to demonstrate? Their walls and valves are weak.

How are they demonstrated? By injecting them with mercury.

Have they been discovered in all tissue? No.

How do lymphatics arise? In lymph spaces.

What do you mean by lymph spaces? Spaces between the fibres.

What is their other name? Juice-canals.

How is nutrient matter applied to the parts? By transudation through the walls of the capillaries.

How does effete matter pass off? Carried by lymphatics from the lymph spaces to the veins.

What tissues receive nutriment through lymph spaces? Cornea of the eye and fibrous membranes.

How does the fluid differ from lymph? It has no lymph corpuscles.

Relation of lymphatics to blood vessels? They are in very close relation.

Do lymphatics vary much in size? No.

What is the size of the smallest that have been demonstrated? One three-hundredth inch.

How do they differ from blood vessels? Their walls are thinner.

Which lymphatics are most numerous? The smallest.

How are they arranged in the skin? Into a superficial and deep plexus.

Which plexus has the largest lymphatics? The deep one.

What are deep lymphatics? Those accompanying deep veins.

What lymphatics have no valves? The smallest.

What is the cause of the beaded appearance of lymphatics? The valves.

What is the course of lymphatics? Generally direct.

How do they anastomose? By bifurcation.

Lymphatics resemble what vessels? Veins.

What are the largest that pass from the skin? One twenty-fifth to one-twelfth inch.

On what does the caliber of the vessel depend? On the pressure of the contained fluid.

Where are lymphatics most numerous on the skin?
Scalp, biparietal suture, palms of the hands and soles of the feet.

Where do they attain their highest degree of development? In the median line of the scrotum.

Give the arrangement of lymphatics on mucous membranes? A double plexus, as in the skin.

In what mucous membrane are they most numerous? Mouth, lips, nares, vagina, penis.

Give their origin and arrangement in the lungs? They arise in the walls of the air cells and form a plexus around each lobule.

In what tissue are they most numerous? In the glandular system.

What peculiarity have they in the brain and cord? They are large and surround the blood vessels.

The vessels on the right side of the head and face empty into what? Into the ductus lymphaticus dexter.

How long is it? One inch.

What is its diameter? One-twelfth to one-eighth inch.

Where does it empty? At the junction of the right subclavian with the internal jugular vein.

Where do the vessels from the inferior extremity empty? Into the thoracic duct.

What are lacteals? Intestinal lymphatics.

Where do they empty? Into the receptaculum chyli.

Where do you find it? Opposite the second lumbar vertebra.

What begins there? The thoracic duct.

Into what does the thoracic duct empty? Into the junction of the left internal jugular with the subclavian vein.

What is the only direct communication between the lymphatics and veins? Where these two ducts empty.

What prevents the blood from passing into the ducts? The valves.

Name the coats of lymphatics? External, middle and internal.

Which are the same as those of the veins? External and internal.

Describe the middle coat? Fibrous and elastic tissue with non-striated muscular fibres.

Why are lymphatics so easily broken? On account of their thin walls.

Where do the valves begin in lymphatics? Near their origin.

How are they arranged? In pairs.

They are most abundant in what lymphatics? In the superficial.

What is the distance between the valves? One-twelfth to one-eighth inch.

Valves are most numerous in what extremity? Lower.

What is their function? To aid the forward flow.

Prove that lymphatics are contractile? They will contract if the thoracic duct is irritated.

Have stomata been demonstrated? No.

Why do you think they exist? The absorption of fats can be accounted for in no other way.

How many lymphatic glands are there? Six hundred to seven hundred.

How divided? Superficial and deep set.

How are the superficial arranged? At the folds and flexures and around the vessels coming from the head.

Where are the deep ones most numerous? Around the vessels coming from the glandular system.

Where do you find the mesenteric glands? Between the folds of the mesentery.

What shape are the glands? Flattened lenticular.

What color are they? Grayish white or reddish.

Consistence of them? Consistence of liver.

Where do the blood-vessels enter? At the hilum.

Where do the efferent vessels emerge? At the hilum.

What is the appearance on the surface? Tuberculated.

Why? The follicles project beneath the investing membrane.

How is the gland divided? Cortical and medullary substance.

What is the color of each? Cortical is reddish gray and the medullary is white.

Give the anatomy of the gland? It is divided into numerous little glands by fibrous tissue.

What are vasa afferentia? Arteries going to the gland.

How many to each gland? Two to six.

Describe them? Pass to the medullary portion and break up into several branches to be distributed to the cortical substance.

How many vasa efferentia? One to three.

Describe them? They are formed from capillary loops, which all center into one or three veins.

How many arteries go to a gland? Two to six.

Where do they penetrate? At the hilum.

What is the action of the lymphatic glands as regards the rapidity of the circulation? They slow it.

What is the function of lymphatic glands? To manufacture white corpuscles.

What is the function of lacteals? To absorb fats.

What else is absorbed by them? Albuminose, saccharine matters and inorganic salts.

Proof of absorption taking place by the skin? Can get physiological effects from medicines given by inunction.

What is the quickest way to produce salivation? By inunctions.

What can be absorbed by the respiratory passages? Gases.

Give an illustration of absorption taking place from a closed cavity? Effusions into serous cavities are gotten rid of in this way.

What change does bile undergo if it remains in the gall bladder? It becomes inspissated.

How are fats absorbed? As an emulsion.

How is mercury absorbed? As such.

Give the variations of absorption? Curara.

How is absorption influenced by hæmorrhage? It increases absorption.

How does the rapidity of the circulation influence it? It increases it.

How does the nervous system influence it? By influencing the blood supply.

How is it influenced by cutting the cerebro-spinal nerve to the blood-vessels? It lessens absorption.

How is the exudation in the lungs in pneumonia gotten rid of? By absorption.

Why do you give digitalis? To increase the rapidity of the circulation.

Why is one thirsty after hæmorrhage? On account of the blood taking water from the tissues.

How are effusions gotten rid of? By tapping and by absorption.

Getting rid of them by absorption is based on what principle? An effort of the blood to remain the same.

What class of medicines are indicated? Diaphoretics, diuretics and hydrogogue cathartics.

CHAPTER XXI.

IMBIBITION.

Define it? *Ans.* Drinking up.

What will influence it? The kind of membrane and liquid and the temperature.

What liquid is imbibed most freely? Distilled water.

What is the effect of alcohol and saline solutions? Retards imbibition.

What membranes imbibe most freely? Homogeneous membranes.

Illustrate the influence of temperature? Put a cow-hide in hot water and one in cold, the former gets soft first.

Describe an endosmometer? A bell glass over which an animal membrane is stretched is floated on water.

Define osmosis? Passing through a membrane.

Define endosmosis? That is the strongest current.

Define exosmosis? The weaker current.

What conditions are necessary for osmosis to take place? There must be two liquids separated by a membrane, the liquids must be mixable and capable of wetting the membrane.

Is it necessary for the liquids to be of different densities? No.

Illustrate the physiological application of these conditions? The products of digestion and the plasma of the blood both wet the wall of the blood vessels and then pass to the blood.

What will produce the most powerful endosmotic current? Albumin.

Does it pass through? It does not.

What change must it undergo to pass through? Must be changed to albuminoids.

Give an illustration of albumin producing a current and not passing through? Break the hard shell from an egg and insert this end into water, in the other end insert a glass tube and the water rises in the egg, forcing the egg into the tube.

Through what kind of a membrane does absorption take place? Homogeneous.

What do you mean by homogeneous? Alike throughout.

What do you mean by heterogeneous? Where it is not the same throughout.

Through what blood vessels does the greater part of absorption take place? The venous radicals.

What is hygrometricity? The power to take on and give off water.

What is water of composition? Water necessary to the make up of a part.

What is interstitial water? Water in the interstices.

Where is the water of composition in a homogeneous membrane? Diffused throughout.

Give the mechanism of osmosis in the case of two liquids that will diffuse with each other and separated by a membrane? Both wet the membrane, one wetting it more than the other, here they mix and then pass to the side of the weakest current.

Give the law of diffusibility of liquids. They diffuse with inverse ratio to their densities.

What effect has chloride of sodium on the diffusion of liquids? Lessens the diffusibility.

What effect has high temperature? Increases it.

Explain the absorption of a saturated solution of cane sugar in the small intestines? The sugar and serum both wet the membrane and mix, they then pass to the exosmotic side, the blood vessels.

Why does the tissue near the gall bladder become stained after death? There is no blood to wash it away.

Give the variation of absorption. Curare.

CHAPTER XXII.

LYMPH AND CHYLE.

What is lymph? *Ans.* The contents of lymphatics and of the thoracic duct during the interval of digestion.

How is it obtained? From the thoracic duct during the intervals of digestion.

How much is there? Four and a half pounds.

Give the properties of lymph? Transparent, yellowish fluid of a characteristic odor, specific gravity 1022.

What does it do if it is drawn and allowed to stand? It clots.

How does the clot differ from that of blood? It is softer and no serum separates it.

Why? Less fibrin produced.

Does it ever coagulate in the vessels during life? Rarely.

Give the composition of lymph? Same as blood with the exception of red corpuscles.

What is the origin of the red corpuscles when you find them? From the blood.

Why is a lymph clot smaller than blood clot? Lymph has less solid matter.

What are lymph corpuscles? Same as white blood corpuscles.

What is the origin of lymph? From the blood and the tissues.

Why think part comes from the tissue? Because lymph has more urea than blood.

What is the function of lymph? To carry excrementitious products from the tissue to the blood.

What is chyle? The contents of the thoracic duct during or just after digestion.

Give its properties? Milky fluid, saline taste, odor of semen, specific gravity 1024.

How is it obtained? Make a fistula into the thoracic duct during digestion.

Quantity? Varies with the kind and quantity of food.

How does it differ from lymph in its composition? It has more albumin, fibrin and salts.

What is the microscopical appearance of chyle? Has a milky appearance.

Has lymph a circulation? Yes.

What kind of a flow? Slow and irregular.

What is the rapidity of the flow? About an inch per second.

What is the main cause of lymph circulation? The force of osmosis and transudation.

What is transudation? A passing through.

What causes assist lymph circulation? Valves, contraction of the heart, expiration and muscular contraction.

Where are lymphatics contractile? The large ones and those of medium size.

How does pulsation of the heart assist? By making pressure on the thoracic duct in its passage under the arch of the aorta.

When does muscular contraction? When the valves are perfect and the contraction intermittent.

What influence has inspiration? Slows it.

Expiration? Increases its rapidity.

CHAPTER XXIII.

SECRETION.

Define it? *Ans.* Something manufactured in a gland.

Do they exist as such in the blood? No.

Where are they produced? In the gland.

When? All the time.

By what is the active principle produced? By the epithelium.

How are the secreting cells divided? Into an outer and inner zone.

Which is the outer zone? The one next to the tubular membrane.

Out of what is the ferment formed? From the ferment forming substance.

What produces pepsinogen? The inner zone of the peptic cells.

What produces pepsin? It is changed pepsinogen.

What is zmogen? Name given to substances which are changed during the digestive ferment.

What is the relation of the blood supply to the discharge of secretions? Direct.

How is the blood supply to a gland at the time of its activity? It is very much increased.

Why? To furnish the watery element to wash out the active principle.

How is it during the interval of activity? Just enough for nutrition.

What part is being formed then? Active principle.

What influence has the composition of the blood on secretion? None.

What effect has blood pressure? The secretion increases with the pressure.

What effect has the nervous system? It influences the blood supply.

Has the nervous system any other influence? It has.

Name it? Mental emotion, pain, etc., cause the flow of tears.

What condition in a gland is necessary for secretion to take place? Must have epithelium.

Give the exception? The mammary gland.

What do you mean by paralytic secretion? Secretion without blood supply.

Give the anatomical classification of glands? Secreting membranes, follicular, tubular racemose and ductless glands.

What is the simplest form? The secreting membrane.

Name one? Synovial membranes.

What do you mean by blood glands? Ductless glands.

How are secretions classified? Transitory and permanent secretions.

Where do you find true synovial membrane? At the diarthrodial articulations.

Give their physiological anatomy? Fibrous and inelastic tissue lined by epithelium.

Describe a bursae? A closed sac with semi-fluid contents placed between a tendon and bone.

Function? To prevent friction.

What is the arrangement of synovial membranes? Around the joint.

Describe the synovial fluid? Colorless, very viscid alkaline fluid.

Give its composition? Water, organic matter, chlorides, phosphates and carbonates.

Function? To lubricate the joint.

Give the anatomical division of mucous membranes? Those covered with pavement and those covered with columnar epithelium.

Name the membranes covered with pavement epithelium? Mouth, lower part of pharynx, œsophagus, conjunctiva, female urethra and vagina.

Give their physiological anatomy? Chorion, basement membrane and epithelium.

Where are the glands situated? In the sub-mucous tissue.

What kind of glands? Simple racemose glands.

Name the membranes covered with columnar epithelium? The alimentary canal below the œsophagus, the ducts of all glands, Eustachian tube, male urethra and Fallopian tube.

What is the condition of the cilia during life? They are in constant motion.

How are they directed? Toward the opening.

Give the physiological anatomy of these membranes? They have a chorion, basement membrane and epithelium.

What kind of glands find in these? Follicular.

What mucous membranes can not be classed with these two varieties? Urinary bladder, pelvis of the kidney and uterus.

Why? They have mixed epithelium.

By what is the mucus produced? Mucous glands.

Give the properties of mucus? Thick, grayish fluid of alkaline reaction.

Where is it not alkaline in reaction? In the vagina.

Why is this acid? To prevent coagulation of the menstrual flow.

What is the action of water on mucus? It swells it.

Exception? The mucus of the conjunctiva is coagulated by water.

Give the composition of mucus? Mucin, water and inorganic salts.

Function of mucus? To lubricate and protect.

Where do you find the sebaceous glands? All parts of the skin where there is hair.

Where are they absent? Palms of the hands and soles of the feet.

What is their relation to the hairs? They are in close relation and they open into the follicle.

How are they divided? Into large and small variety.

Where do you find the small ones? In relation with the large hairs.

Where find the large ones? In relation with the small hairs.

What kind of glands are they? Compound or simple racemose.

How many empty into a large hair follicle? Four to eight.

Where are the largest glands found? Nose, eye, ear, penis and areola of the nipple.

Where do you find the glands of Tyson? They are the sebaceous glands of the glans penis.

Give their physiological anatomy? Same as other compound racemose glands.

Where do you find the ceruminous glands? In the cartilaginous part of the ear.

Describe them? A tube which terminates in a coil.

Where do you find the Meibomian glands? In the tarsal cartilages.

Describe them? Made up of a straight duct into which a number of compound racemose glands open.

Give the composition of sebaceous matter? Albumin, fats, chlorides, phosphates.

What is smegma? Sebaceous matter of the glans penis in the male and labia minora in female.

What is vernix caseosa? Sebaceous matter covering the fœtus at birth.

What is the function of it? To protect the fœtus.

What is the function of sebaceous matter? To lubricate the hair and to protect.

What is cerumen? Secretion of the ceruminous glands.

What is its function? To keep foreign bodies out of the ear.

On what does its color depend? On the time it has been exposed to the air.

What is the function of meobomian secretion? To prevent an overflow of tears.

CHAPTER XXIV.

MAMMARY GLANDS.

What is their shape in the female? *Ans.* Hemispherical.

Where are they situated? Anterior part of the thorax over the pectoral muscles.

From what point do they extend vertically? Second to the seventh rib.

What is their extent laterally? From axilla to sternum.

How are they before puberty? Undeveloped.

How are they during pregnancy? They become enlarged.

What effect has old age? They atrophy.

Which gland is the largest? The left one.

What separates them from the pectoral muscles? Deep fascia.

What is the shape of the outer surface? Convex.

What is the maurilla? The nipple.

Give their anatomy? Ducts, muscular and connective tissue and blood vessels.

What is the arrangement of the muscular fibres? They run in no definite direction.

What is the color of the surface? Brown.

By what is it surrounded? By the areola.

What color is the areola in a virgin? Pink.

What influence has pregnancy? It darkens it.

What is the color after lactation? Dark brown.

What is the function of the muscular fibres? To compress the milk ducts.

What is the shape of the nipple? Conical.

What effect has mechanical excitement? Makes it erect.

Why? It causes the muscular fibres to contract.

What do you find on the surface of the nipple? Papillæ.

What is the cause of the tuberculated appearance of the areola? It is caused by the large sebaceous glands found beneath it.

On what does the areola rest? On the subjacent glandular structure.

What is the function of these glands? To moisten.

To what class of glands do the mammary glands belong? Compound racemose.

Of what are they made up? Lobes.

How many lobes? Fifteen to twenty-four.

How are the lobes divided? Into lobules.

Of what does the smallest lobule consist? Secreting vesicles.

How many excretory ducts are there? Ten to fourteen.

What are they called? Lactiferous or galactophorous ducts.

Where are they dilated? Beneath the areola.

What is it called? Galactophorous sinuses.

What is the function of these dilatations? They are receptacles for the milk.

9

How much do they hold? About two fluid ounces.

Name the coats of the ducts? External, middle and internal.

Describe the external? Fibrous tissue with some elastic fibres.

Describe the middle? Non-striated muscular fibres arranged longitudinally.

What is the function of this coat? To shorten the duct.

Describe the internal? It is simply an amorphous membrane.

Of what is each acinus made up? Twenty to forty secreting vesicles.

What lines these vesicles? Epithelium.

When? During the interval of lactation.

What kind of epithelium? Polygonal cells.

What becomes of it during lactation? It is exfoliated.

What is the condition of the ducts during the interval of lactation? They become retracted.

Where are the important and characteristic constituents of milk formed? In the gland.

What is the origin of the inorganic principles? They are separated from the blood.

Name the constituents produced in the gland? Sugar, fat, and casein.

Is milk a secretion or excretion? Secretion.

When is milk produced? All the time.

Does the quantity of food influence the quality or quantity of milk? It does not.

What effect have liquids? It increases the amount.

Alcohol? Increases it.

Mental emotion? Lessens it, or it may stop it.

How much milk is produced in a day? Two to three pints.

How much can be drawn from a full breast? Two fluid ounces.

What effect has the menstrual period on the amount of milk? It diminishes it.

Pregnancy? It also diminishes it.

Give the properties of milk? White, opaque, sweet fluid, alkaline in reaction, specific gravity 1032.

What do you mean by coagulation? Separation into whey and curd.

What effect has a thunder storm on milk? It coagulates it.

If milk stands how does it separate? Into milk and cream.

How much cream should there be in healthy milk? One-fifth to one-third.

Give the composition of cream? Almost entirely milk globules.

What is the specific gravity of cream? 1024.

What is the microscopical appearance of milk? Minute globules held in suspension in a clear fluid.

What are the globules called? Milk globules.

Give the composition of milk? Water, casein, albumin and inorganic salts.

How much water? Nine hundred parts.

What is the name for the nitrogenized constituent? Casein.

Where is it produced? In the gland.

From what? Principles derived from the blood.

In what form is it in milk? In solution.

How is it when first drawn? In solution.

How does the milk of a cow differ from mother's milk? Mother's milk is poorer in casein and richer in butter and sugar.

What is the object of feeding a child on cow's milk? To make it as near as possible like the mother's milk.

How make cow's milk like a woman's? Take the cream and add water and sugar.

On what does the coagulation of milk depend? On the formation of lactic acid.

How is lactic acid formed? By the decomposition of sugar of milk.

What holds the fats in milk in emulsion? The liquid casein and water.

What is the other name for the fats? Butter.

How does butter from a woman's milk differ from that of a cow? It is softer.

Give the composition of butter? Olein, palmitin, capriline, etc.

What is lactose? Sugar of milk.

Give the peculiarities of milk sugar? It is readily changed to lactic acid, but takes on alcoholic fermentation slowly.

What is the function of the alkaline carbonates? To preserve the fluidity of the casein.

In a general way what does milk contain? All the elements necessary for nutrition.

Are there any variations in its composition? Yes.

Name them? Milk during lactation and pregnancy and colostrum.

How is the child affected by the milk when menstruation begins? It becomes emaciated.

How, by pregnancy? It becomes emaciated.

Why? Milk then has little nutritious properties.

What is colostrum? The secretion of the mammary glands just after parturition.

Give its composition? It has more sugar and inorganic salts than milk.

Its function? To relieve the infant of the accummulated meconium.

What causes it to act as a laxative ? The sugar and inorganic salts.

What is meconium? The first fæcal matter of the child.

How long does milk preserve the characters of colostrum ? Eight to ten days.

Give the composition of the milk of an infant? Less butter and more sugar than woman's milk.

CHAPTER XXV.

EXCRETION.

Define it? *Ans.* Something thrown off from the body.

What fluids carry it from the tissue? Blood and lymph.

Where is it carried? To the organs of excretion.

What becomes of them ? They are thrown off.

Do they in any way aid in digestion? Only by getting rid of the waste products.

Are they produced in glands? No.

What trouble if they are not thrown off? The system becomes poisoned.

How are they discharged? Continuously.

Why? To prevent poisoning of the system.

What excrementitious product does nervous tissue produce when it wears out? Cholesterin.

Where is it gotten rid of? By the liver.

What does the wear of muscular tissue produce? Urea.

How eliminated ? By the kidney and skin.

CHAPTER XXVI.

SKIN.

Name the function of the skin? *Ans.* To preserve the form of the muscles, and it acts as an organ of excretion and sensation and as a protectile covering.

What is the extent of the skin? Sixteen square feet.

Where is it thickest? Where it is most exposed.

What is the thickness of the skin? One-half to one-eighth inch.

Where is it thinnest? Glans penis, lips and external ear.

How is the skin naturally divided? Into two layers.

Names? Epidermis, or false skin, and dermis, or true.

What is the panniculus adiposus? The layer of fat beneath the skin.

Where is it absent? Outer ear, eyelids and penis.

Where is it thickest? On the abdomen.

What is the condition of the under surface of the skin? It is rough.

How is the true skin divided? Into a reticulated and papillary layer.

Give the anatomy of the reticulated layer? Elastic and fibrous tissue with some non-striated muscular fibres.

Where are the muscular fibres most abundant? Around the sebaceous glands and hair follicles.

Explain "goose flesh"? It is caused by a projection of the follicles above the skin.

Where is the contraction most marked? Nipple, penis and scrotum.

Describe the papillary layer? A simple amorphous membrane.

How are the papillæ formed? By eversions of this layer.

How are they divided? Into simple and compound papillæ.

Where are they largest? Nipple, palms of the hands and soles of the feet.

How do the blood vessels terminate in the papillæ? They form a flexus.

What is the epidermis? The false or scarf skin.

What kind of tissue is it? Extra vascular tissue.

Name the layers that make it up? Malpighian and horny layer.

Give the composition of the Malpighian layer? A single layer of primordial cells with coloring matter.

On what does the color depend? On the amount and tint of the pigmentary matter in the Malpighian layer.

What is the thickness of this layer? One seven-thousandth to one seventy-fifth inch.

Describe the horny layer? Several strata of hard flattened cells.

What is the function of the epidermis? It gives protection.

Where is it thickest? Palms of the hands and soles of the feet.

What becomes of the epidermis when it wears out? It is thrown off.

Name the appendages of the skin? Nails and hair.

How is a nail divided? Into root, body and free extremity.

Describe the root? Its edge is thin and ragged, and this part is softer than the rest of the nail.

How long is the root? One-fourth to one-third the entire length.

Describe the body? Thick and is marked by longitudinal lines; it extends from root to free extremity.

What is the lunula? The semilunar white mark on the body near the root.

What causes it? This part is less vascular.

What is the matrix? The part of the skin on which the nail rest.

What is the free border? That extends beyond the matrix.

How long may it become? Two inches.

Where is the nail thickest? Central portion.

How many layers has a nail? Two.

How does the nail grow? The horny layer grows from the bed and from the root.

What are the functions of the nail? Organs of prehension and protection.

On what parts of the body are hairs not found? Soles of the feet. palms of the hands, outer ear, penis, labia minora, etc.

How are hairs divided? Long, short and downy hairs.

Where are the long hairs found? Head, axilla, and around the genitals.

Where do you find the short hairs? Around the openings.

Where are the downy hairs found? Over the general surface.

How are hairs set in the skin? Obliquely.

How long do long hairs become? Two or three feet.

How long are the short? One-fourth to one-half inch.

How do hairs terminate that have never been cut? In a pointed extremity.

Is hair coarser in male or female? Female.

Is it coarser in blondes or brunettes? In the brunettes.

Give the properties of hair? Elastic, hygrometric, and it becomes electric on friction.

What is the root of the hair? That part imbedded in the skin.

How deep do the roots penetrate? The downy go only to the true skin and the large go through it.

Describe the root? Soft, large and terminates in a bulb.

What surrounds the root? The sheath.

What surrounds the sheath? The hair follicle.

What forms the sheath? The horny layer of the skin.

What forms the follicle? The malpighian layer.

Give the structure of hair? They are fibrous tissue and it is covered by small cells.

How is the hair nourished? By imbibition.

On what does the color of hair depend? On the pigmentary matter.

What is the cause of sudden bleaching? The medulla fills with air.

What is the cause of blanching in old age? The loss of pigmentary matter.

CHAPTER XXVII.

PERSPIRATION.

Define it? *Ans.* The action of the skin as an excretory organ.

Is the skin an organ of excretion or secretion? Excretion.

Name the products excreted by the skin? Urea, carbonic acid, water and organic matter.

What would be the result of covering the skin with some impermeable substance? Would result in death.

What would be the cause of death? It would suppress the action of the skin and depress the temperature.

What glands in the skin are most numerous? Sudoriparous glands.

Where do you not find them? Concha of the ear, penis, prepuce, and external ear.

Where are the orifices of the ducts in the palms of the hands and soles of the feet? Between the papillary ridges.

How do the tubes open on the surface? Obliquely.

Describe a sudoriparous gland? A simple tube terminating in a coil.

What is the size of the coil? One one-hundred and twenty-fifth to one twenty-fifth of an inch.

Where are they smallest? Eyelids, ear, nose, penis and scrotum.

Where are they largest? Areola of the nipple and the perineum.

How many convolutions in the coiled portion? Six to twelve.

Describe the blood-vessels around the coil? They form a plexus around it and send branches between the convolutions.

What is the excretory duct? A continuation of the coil.

How does it pass through the true skin? In a straight course.

How through the false skin? Spiral course.

Where is the excretory duct largest? Near the opening.

What is the peculiarity of the glands of the axilla? They are large and excrete sweat of a peculiar odor.

How many sweat glands are there? 2,381,000.

How long is each coil? One-sixteenth inch.

How long are they all together? Two and one-third miles.

Is the action of the skin continuous or intermittent? Continuous.

How is sweat exhaled from the skin? In the form of vapor.

What part of the gland performs the function of excretion? The epithelium of the coil.

How is perspiration excreted? Continuously.

What influence has the nervous system on the production of sweat? It influences it by influencing the blood supply to the glands.

Where is the sweat center located? In the lower part of the cervical or dorsal regions.

What effect would cutting of the sympathetic nerve going to the blood-vessels of the skin have? It would increase the amount of sweat.

What is the amount of cutaneous exhalation? About two pounds per day.

What time of the year is it greatest? Summer.

What varies least in sweat? Urea.

Give the properties of sweat? Colorless limpid liquid, of acid reaction and characteristic odor.

Why does sweat become alkaline? The volatile fatty acids escape.

What is the specific gravity of sweat? 1003 to 1004.

Give its composition? Water, urea, fatty matter, chlorides, phosphates and sulphates.

What is the most important part of sweat? Urea.

What kind of acids are excreted? Fatty acids.

Name some of them? Formic, butyric, cupric and acetic acids.

What parts produce sweat with peculiarities? Axilla, between the toes and around the genitals.

Describe the sweat produced by them? It is alkaline in reaction, has a strong odor and stains yellow.

What is the cause of its odor? It is caused by the secretions from other follicles in these parts.

CHAPTER XXVIII.

KIDNEYS.

What was the old idea of the formation of urea? *Ins.* That it was formed in the kidneys.

Give the reasons why they thought so? (1) When the kidneys were removed there was no uremia for several days. (2) When urea was injected into the veins there was no uremia. (3) When the ureters were tied there would be uremia.

Where is urea produced? In the tissue.

It represents the wear of what tissue? Muscular tissue.

Why is there not an increased quantity of urea in the blood the first few days after removal of the kidneys? It is eliminated vicariously.

What organs normally eliminate urea? Skin and kidney.

Which eliminate it vicariously? Lungs and alimentary canal.

What change does urea undergo in contact with a mucous surface? Changes to carbonate of ammonia.

What symptom would it produce in the bowel? Diarrhœa.

What in the stomach? Vomiting.

In the lungs? Coughing.

Why does it produce these symptoms? It acts as an irritant.

When urea is injected into the blood, why will there be no uremia? It acts as a diuretic, which increases the amount eliminated by the kidneys.

Why will tying of the ureters increase the quantity of urea in the blood, while cutting the kidneys out will not for several days? In tying the ureters the pain prevents the vicarious elimination of urea.

What takes the urea from the tissues? The blood and lymph.

Where is it carried? To the skin and kidneys.

What is the color of the blood in the renal vein? Bright red.

Why? Has a large amount of oxygen.

What is the purest blood in the body? That of the renal vein.

Why? It has no urea and has much oxygen.

Is urea an excretion or secretion? Excretion.

What part of the kidney separates it from the blood? The epithelium in the contorted tube.

In Bright's disease what becomes of the epithelium? It is exfoliated.

Is it reproduced? It is not.

How is the urea eliminated when the epithelium is destroyed? Vicariously.

Where is the watery part of the urine separated from the blood? In the Malpighian corpuscle.

What is the effect of extirpating one kidney? It makes the other one act continuously.

What is the effect of extirpating a kidney on the appetite? It increases it.

What influence has the blood pressure on the quantity of urine? The higher the pressure the greater is the amount of urine.

Why will digitalis act as a diuretic? It increases the blood pressure.

Effect of cutting the nerve going to the kidney? It increases the flow of urine.

Effect of injuring the pneumogastric nerve near the origin on the quantity of urine? Increases it.

Effect of injuring the floor of the fourth ventricle? Would cause diabetes mellitis.

What effect has digestion on the amount of urine? It increases it.

When is urine formed? All the time.

Why? To get rid of urea.

Do both kidneys act at once? They do not.

What change is there in the composition of the blood in passing through the kidney? It loses urea and water.

What is the upper expansion of the ureters called? The pelvis of the kidney.

What is the calyx? A process of the pelvis.

What do the calyces receive? The apices of the pyramids.

What is the diameter of the ureters? About same as that of a goose quill.

How long are they? Fifteen to eighteen inches.

What is their extent? From kidney to bladder.

How many coats have they? Three.

Describe the external coat? Fibrous tissue with some elastic fibres.

Describe the middle coat? Non-striated muscular fibres arranged in no definite direction.

Describe the internal coat? It is a thin, smooth, mucous membrane.

Where are the ureters constricted? Near the bladder.

How do they penetrate the bladder? Obliquely.

What is the object of this? To prevent regurgitation of urine.

What is the function of the bladder? Acts as a reservoir for the urine.

Where do the ureters empty? At the base of the trigonum.

Where do you find it? At the base of the bladder.

What is the shape of the bladder? Triangular.

What shape is it when distended? Globular.

Where do you find the bladder? In the pelvic cavity.

How much will the bladder hold? Normally from one to two pints.

Is it largest in male or female ? Female.

How is the bladder held in place? By ligaments.

Name the coats of the bladder? External, middle and internal.

Describe the external coat ? It is peritoneal and is complete only behind.

How many layers in the middle coat ? Three.

Describe the internal? Longitudinal mucular fibres running antero-posteriorly.

Describe the middle layer of this coat ? Circular fibres, arranged in bands on the anterior surface.

Describe the internal layer? Pale fibres arranged longitudinally.

With what are they continuous? With the muscular fibres of the ureters, urachus and urethra.

Describe the sphincter vesicæ ? A collection of circular fibres a half inch wide, found at the neck of the bladder.

What is its function? To prevent a flow of urine and also to keep the semen out of the bladder.

Describe the mucous membrane of the bladder? Smooth, pale, thick and loosely attached to the sub-mucous tissue, except at the corpus trigonum.

What is the shape of the trigonum? Triangular.

What is its composition? Fibrous tissue with some muscular fibres and elastic tissue.

What is the uvula vesicæ ? A fold of the mucous membrane at the urethral opening.

What part of the mucous membrane are the blood-vessels most numerous? At the fundus.

What is the cause of the discharge of urine into the pelvis of the kidney? The pressure brought about by separation of the urine from the blood.

What causes it to flow through the ureter? Force from behind and contraction of the ureters.

What causes the desire to void urine? Pressure on the sensory nerves of the bladder.

How frequent is it? Three to five times per day.

Give the mechanism of the discharge from the bladder? The sphincter dilates and the urine is forced out by contraction of the bladder and abdominal muscles.

Is it voluntary? It is so at first.

Where is the center located that is supposed to control the bladder? Opposite the fourth lumbar vertebra.

Give the properties of urine? Clear amber-colored nonviscid fluid with aromatic odor, acid reaction, and specific gravity 1020.

How much urine is passed in a day? About fifty-two ounces.

What is the cause of the acidity of urine? Due to the acid phosphate of soda.

Give the composition of urine? Water, urea, creatin and creatinin carbonates, phosphates, chlorides and sulphates.

Which is the most important constituent? Urea.

Give its source? From the oxidation of the tissue.

Can it be produced in the laboratory? Yes.

How? By the action of sulphate of ammonia on cyanate of potash.

Into what is it converted when it comes in contact with a mucous surface? Carbonate of ammonia.

What is the chemical difference between urea and carbonate of ammonia? The ammonia has four more atoms of water.

Where is urea normally found? Urine, blood, lymph, chyle and sweat.

How can urea be made from carbonate of ammonia? Heat it to drive off the water.

In what form is urea in the urine? In solution.

What effect has watery diet on urea? It increases the amount excreted.

Effect of watery diet on specific gravity? It lowers it.

What effect has nitrogenized diet? Increases the amount of urea.

What effect has non-nitrogenized diet? It lessens it.

Effect of muscular activity? Increases it.

How much urea is eliminated daily? 355 to 460 grains.

Does uric acid exist in a free state in urine? No.

What is its solubility in hot water? 1800 parts.

In cold water? 14000 to 15000 parts.

In what form is uric acid found? In the form of urates.

Which urate is most important? Urate of soda.

What is the origin of uric acid? From the oxidation of tissue.

How much is excreted in a day? Six to ten grains.

In what urine do you find hippuric acid? In the urine of herbivora.

Is it a normal constituent of human urine? It is not.

How much is eliminated per day? Seven and one-half grains.

What is its solubility? Six hundred parts of cold water, more soluble in dilute hydrochloric acid.

What is the source of the lactates? Formed by the action of lactic acid on the bases.

What is the source of lactic acid? From the transformation of glucose.

What is the source of creatin and creatinin? Waste of tissue.

Higher oxidation of them produces what? Urea.

How much creatin and creatinin eliminated in a day? Eleven and one-half grains.

Does oxalate of lime exist in normal urine? None constantly.

10

What is oxalyuria? An excess of oxalates in the urine.

What is the origin of the inorganic salts in the urine? Come from the tissues.

What is the proportion of chloride of sodium in the blood? Three to four parts per thousand.

In the urine? Three to eight parts.

On what does the amount in the urine depend? On the amount taken in.

What is the relation of febrile troubles on the amount of chloride of sodium in the urine? They lessen the amount.

How much is eliminated daily? One hundred and fifty-four grains.

What is the proportion of sulphates in the urine? Three to seven parts per thousand.

What is their origin? From within and without.

What was the old significance of phosphates in the urine? That it was due to brain activity.

Why? Because nervous tissue is made up largely of phosphates.

What is the origin of the phosphates? From without.

Name the different kinds? Neutral, basic and acid phosphates.

Which is the most important one? Acid phosphate of sodium.

What is the coloring matter of the urine? Urochrome.

What is its origin? It is probably formed from hæmoglobin.

Name the gases in the urine? Oxygen, nitrogen, and carbonic acid gas.

What becomes of the hydrogen introduced with the food? It goes to form water.

Proof of water being formed in the body? There is more discharged than taken in.

At what period of life is the most urine discharged in proportion to the size of the body? Infancy.

When is the most urea discharged? Infancy.

Is there more urine discharged in male or female? Male.

What time of the year is most urine discharged? Winter.

Why is there less in summer? It is mostly eliminated by the skin.

When is the specific gravity of the urine highest? In the morning.

Why? Because it has not been voided during the night.

What is the color of urine passed in the forenoon? Light.

How is the specific gravity? Low.

What influence has nitrogenized diet on urine? It increases the amount of urea and increases its specific gravity.

Effect of liquid food and alcoholic drinks? Lowers its specific gravity.

Effect of coffee and tea? Lowers it and lessens the urea.

What effect has muscular exercise on the amount of urea? Increases it.

Why? It causes more waste of tissue.

Effect of mental exercise? Increases it.

Why? It causes waste of tissue.

CHAPTER XXIX.

LIVER.

What part of the liver performs the office of secretion and excretion? *Ans.* The hepatic cells.

State the functions of the liver? Secretion, excretion, and the function of a ductless gland.

What part is an excretion? Cholesterin.

What part is a secretion? Biliary salts and coloring matter.

What is the office of a ductless gland? To manufacture something and discharge it into the blood.

What was formerly thought to manufacture the bile? The racemose glands opening into the biliary duct.

What is the function of these racemose glands? To produce mucus.

Is bile secreted from arterial or venous blood? Mostly from venous.

What are the branches of the hepatic artery in the liver? Vaginal and interlobular.

What are the branches of the hepatic vein? Intralobular and sub-lobular.

Of the portal vein? Vaginal, interlobular and lobular.

How does the blood get from the liver to the inferior vena cava? Through the hepatic vein.

Prove that bile can be secreted from arterial blood? Tie the portal vein and will still have bile produced.

How much bile is secreted in a day? Two to four pounds.

When is bile discharged into the duodenum? All the time.

When is the greatest quantity discharged? During digestion.

Why is it necessary for bile to be discharged all the time? To get rid of cholesterin.

Give the functions of bile? Nature's purge and disinfectant, and to aid in digestion.

What kind of food does it help digest? Fats.

Is bile essential to life? It is.

Proof? Make a biliary fistula and will result in death.

Give the properties of bile? Thick, golden green opaque fluid of a bitter taste and specific gravity 1020 to 1026.

Composition? Water, biliary salts, cholesterin, inorganic salts and coloring matter.

What is the origin of the inorganic salts? From without.

What is their function? They seem to be of little use.

What is lecithene? A neutral fatty substance found in bile.

Where is it found? Blood, bile, nervous tissue, and yolk of egg.

What is meant by the recrementitious function of bile? Its function in digestion.

What part of it has to do with this function? The biliary salts.

Name the biliary salts? Taurocholate and glycocholate of soda.

How are they formed? By the action of taurocholic and glycocholic acid on soda.

Which is found in greatest abundance? Taurocholate.

What is the difference between the two? Sulphate of aluminum and iron precipitates taurocholate and not the other.

Are they proximate principles of bile? No.

What is their origin? They are formed in the liver.

Name the excrementitious principles of bile? Cholesterin.

How is it held in solution in the blood? By the fatty acids.

How is it held in solution in the bile? By the biliary salts.

What are biliary calculi? A collection of crystals of cholesterin.

In what form is cholesterin usually found? In solution.

Name the coloring matter of bile? Bilirubin.

What is its origin? It is formed in the liver, probably from hæmoglobin.

Give the test for it? With nitroso nitric acid it gives a play of colors.

Give the test for the biliary salts? Gives a purple with cane sugar and sulphuric acid.

What is the excrementitious function of the liver? To excrete cholesterin.

What is cholesterin? A product of waste of nervous tissue.

Proof? There is more in the blood coming from nervous tissue.

Is there any formed in the lower extremity? There is.

Proof of its elimination by the liver? If the function of the liver is stopped the system becomes poisoned with cholesterin.

What is stercorine? Modified cholesterin.

When is it discharged as cholesterin? During the intervals of digestions.

What office of the liver resembles that of a ductless gland? The glycogenic action.

What does the liver constantly contain? Glycogen.

What becomes of the glycogen? It is converted into glucose.

What is the composition of glycogen? Carbon, hydrogen and oxygen.

It belongs to what class? Carbo hydrates.

Give the test for glycogen? With iodine it gives a deep red.
What does it resemble? Starch.

How does it exist in the liver cells? In amorphous granules.

How can you extract it from a decoction? By precipitating it with alcohol.

Where do you find it in fœtal life? Through all the tissue.

What is the effect of digestion on glycogen? It increases the amount.

What kind of diet will diminish it? Albuminous and fatty diet.

What will increase it? Carbo hydrates.

What change does sugar undergo in the liver? It is changed to glucose.

What is the source of glycogen in herbivora and in man? From the starch in the vegetable food.

Give evidence of the glycogenic function of the liver? Sugar is found coming from but not going to the liver.

When is sugar found in the portal vein? When an excess is taken in.

When in the hepatic artery? When an excess is taken in.

Give a test for sugar? Boil with caustic potash and it gets yellow or brown, as to the amount of sugar.

What kind of blood goes from the liver? Venous.

In what blood is sugar found? Blood from the liver to the lungs.

What becomes of it? It is destroyed in the lungs.

Is it ever deposited? No.

It is supposed to have to do with what? With the production of fats.

Why think so? Fat and sugar have the same elements, and people using much sugar grow fat.

What elements make up sugar? Carbon, hydrogen and oxygen.

Is sugar normally found in the blood going to the liver ? No.

Is it found in that going from the liver? It is.

What would that indicate? That it is formed in the liver.

Does the liver contain sugar during life ? It does not.

Proof? As soon as an animal is killed, test for the sugar and will not find it.

In what time must it be tested ? In ten seconds.

Give the characters of liver sugar? It ferments more readily than other sugar.

Give the mechanism of the production of sugar in the liver ? The hepatic ferment acts on the glycogen.

What is the first organ to produce sugar ? The placenta.

What influence has digestion on the production of sugar? It increases it.

Has diet any effect ? Yes. Some increase and others lessen it.

What is the destination of sugar ? It is destroyed in the lungs.

What kind of bread do you use in glycosuria ? Bread made of gluten.

Why ? It has a variety of principles.

How is sugar production influenced by cutting the pneumogastric between lungs and liver ? Not affected.

Between the origin and lungs? It is not affected.

Effect of irritation of the fourth ventricle of the brain ? Causes diabetes mellitus.

Effect of stimulating the peripheral extremity of cut nerve in neck ? Has no effect.

Stimulating the central end ? Increases the sugar.

Through what system of nerves is it transmitted ? Sympathetic.

What effect has the inhalation of anæsthetics ? Increases the sugar in the urine.

CHAPTER XXX.

DUCTLESS GLANDS.

Is the spleen essential to life? *Ans.* It is not.

Proof? Can do without it.

What is it thought to have to do with the manufacturing of leucocytes? Leucocytes are found in greater number in the splenic vein than artery.

What is the liver thought to have to do with destruction of red? They are found in diminished quantity in the hepatic vein.

Why is the spleen thought to act as a diverticulum for the blood during digestion? It is found full of blood at this time.

How is the appetite when the spleen is removed? It is very much increased.

In what trouble is the spleen enlarged? Malaria.

Where do you find the suprarenal capsule? Above and in close apposition with the kidney.

Is its function known? It is not.

In what trouble are they diseased? Addison's disease.

Are they essential to life? They are not.

Where do you find the thyroid gland? One lobe on each side of the trachea.

Is it essential to life? No.

Is its function known? No.

What is enlargement of it called? Goitre or bronchocelle.

When is it largest in the female? At the menstrual period.

At what period of life is the thymus gland largest? In foetal life.

CHAPTER XXXI.

What is nutrition? *Ans.* Appropriating new material for the waste of the old.

What becomes of all tissue? It wears out.

What parts when they wear out are not carried by the blood to the organs for elimination? False skin and its appendages.

How are they gotten rid of? They are exfoliated.

What trouble would there be if these effete materials were not eliminated? The system would become poisoned.

What was the first process of nutrition that we studied? Circulation of the blood.

What does the blood contain? All the elements necessary for nutrition.

What was the second process we studied? Respiration.

What is respiration? The process by which the tissues receive oxygen and give off carbonic acid.

What is the relation of respiration to nutrition? Direct.

How long would we live without oxygen? Three to seven minutes.

What is the third act? Alimentation.

What classes of food are essential to life? Organic, nitrogenized and inorganic.

How long would we live without food? About forty days.

What is the fourth act? Deglutition.

What is deglutition? The act of swallowing.

How is it divided? Into three acts.

What is the fifth act? Digestion.

What is digestion? Liquefaction of the food.

Where does it begin? In the mouth.

Where end? In the small intestines.

What is the first object of digestion? Liquefaction of the food.

What is the second object of digestion? To prepare the food for absorption.

By what vessels is absorption carried on? Veins and lymphatics.

What is absorbed by the veins? All that will form a homogeneous mass with the blood.

What by the lymphatics? All the rest.

What is the sixth act of nutrition? Secretion.

What is a secretion? Something manufactured in a gland.

How are they divided? Permanent and transitory secretions. ,

What is the function of the permanent? To lubricate.

Of the transitory? To aid in digestion.

Are they discharged from the organism? No.

What is the seventh act of nutrition? Excretion.

What trouble if excretions are disturbed? The system becomes poisoned.

What is an excretion? Something discharged from the body.

What principles pass through the organism unchanged? The inorganic.

How many are there? Twenty-one.

Name the gases? Oxygen, nitrogen, hydrogen, sulphuretted and carburetted hydrogen.

Where are oxygen and hydrogen found? Everywhere.

Where is nitrogen found? In all tissue.

Where are carburetted hydrogen and sulphuretted hydrogen found? In the alimentary canal.

Which is essential? Oxygen.

What is most imperative after oxygen? Water.

Where do you find it? Everywhere.

What regulates the amount? Chloride of sodium.

How can you tell when there is too little? Become thirsty.

If too much? Have dropsy.

What is the function of water? It acts as a solvent for the salts, gives elasticity to cartilage, contractility to muscle, density to bone and pliability to tendons.

What is the origin of water? From within and without the body.

Why think it is formed in the body? There is more discharged than taken in.

How much is discharged by the kidneys daily? Fifty-two ounces.

By the lungs? One and one-fifth pounds.

How much water is there in the blood? About eight hundred parts.

Where is chloride of sodium found? Everywhere except in the enamel of the teeth.

Is there more in solids or fluids? Fluids.

How much is there in the blood? Three to four parts per thousand.

How much in urine? Three to eight parts.

What is the origin of chloride of sodium? Introduced from without.

How is it discharged? Sweat, urine and mucus.

What is the function of chloride of sodium? To give density to the plasma, aid in osmosis and determine transudation.

What is the origin of chloride of potash? From without and from double decomposition within.

How is it discharged? In mucus and urine.

What is its function? Same as chloride of sodium.

Where is phosphate of lime found? In all the solids and fluids of the body.

Is there more in solids or fluids? Solids.

Is it very soluble? No.

What holds it in solution? The carbonic acid and carbonates.

It is found especially in what tissue? Bony tissue.

In what bones is it in greatest quantity? Those of the lower extremity.

Why? So that they can sustain the weight of the body.

At what period of life is there most? Old age.

Why? There is a deficiency of its elimination at this time.

When are fractures most liable? In old age.

Why? The bones are more brittle.

What is the origin of phosphate of lime? Introduced from without.

How is it eliminated? In the urine and fæces.

What is its function? To give to the bones their power of resistance.

What is the function of the other phosphates? The same.

Where is carbonate of lime found? Blood, bones, teeth, cartilage and internal ear.

How does it differ from the other salts in the organism, in its form? It is found in the crystalline form in the internal ear.

What is its function? Thought to have to do with audition.

What is its origin? Introduced from without and formed within.

How is it discharged? In the urine as phosphate of lime.

Is carbonate of soda ever introduced as such? No.

How is it formed? By the decomposition of tartrates, malates, etc.

What is its function? To preserve the fluidity of the albuminoids and the consistency of the corpuscles.

What is the function of the sulphates? Same as carbonate of soda.

What is the origin of sulphates? Introduced from without.

Name the principles consumed in the organism? The organic.

How are they divided? Into nitrogenized and non-nitrogenized.

Name the nitrogenized class? Albumin, musculin, casein, etc.

What is another name for this class? Albuminoids or plastics.

What elements are in this class? Carbon, hydrogen, oxygen, nitrogen and sulphur.

Do they produce fats? No.

What do they produce? Muscular tissue.

What tissue is especially nourished by nitrogenized diet? Muscular tissue.

How is urea influenced by nitrogenized diet? It is increased in amount.

What is the object in training? To introduce nitrogenized food enough to compensate the waste and to take a moderate amount of exercise.

Power resides in what kind of tissue? Muscular tissue.

What is crystalline? The nitrogenized substance found in the crystalline lens.

What is myosin? Nitrogenized substance of muscles.

What is keratin? That of the epidermis and its appendages.

What is elastin? That of elastic tissue.

Ossein? That of bones.

Gelatine? That of fibrous tissue.

What is meant by metabolism? The changes in the tissue in digestion.

Name the non-nitrogenized food? Sugar, starch and fat.

What is another name for this class? Calorifics.

Are the non-nitrogenized principles essential to life? No.

Why are they called calorifics? They are supposed to have to do with keeping up animal temperature.

Are they essential to normal temperature? No.

When are they most called for? In winter.

When is fat deposited in the tissue? When an excess of food has been taken in.

Is there any power in fatty tissue? No.

If deprived of food, which lives longest, a fat or lean person? Fat one.

What part of the tissue is first used up? The fat.

Is fat formed in the body? It is.

Reason for thinking so? A person can get fat without the introduction of fat.

Give reasons why think sugar produces fat? They contain the same elements, and people using much sugar grow fat.

How is fat found in the body? As adipose tissue.

Give its physiological anatomy? A nitrogenized net work in which are held oil globules.

Are there any nerves or lymphatics in fatty tissue? There are not.

CHAPTER XXXII.

ANIMAL HEAT.

How many heat units are produced by each pound per hour? *Ans.* Four.

What is the normal temperature? Ninety-eight and one-half degrees Fahrenheit.

Give the variations? One-half degree Fahrenheit below and one degree above that.

Give the variations with the external temperature? Eighty three degrees Fahrenheit and one hundred and seven degrees Fahrenheit.

On which side of the heart is the blood warmest? Right.

Why? It has not been cooled in passing through the lungs.

What is the warmest blood in the body? That of the hepatic vein.

What is its temperature? One hundred and seven degrees Fahrenheit.

Is it warmer in the portal vein or abdominal aorta? Portal vein.

What period of our life is the temperature lowest? Just after birth.

At what period do we resist cold least? In early life.

At what period of the day is the temperature highest? At 11 A. M. and 4 P. M.

What effect has fasting on the temperature? It lowers it.

Why? You have defective nutrition at this time.

What effect has exercise on it? It increases it.

What is the source of heat? The oxidation of tissue.

What was the old idea of its cause? That it was produced by some special organ.

What relation has heat to nutrition and to respiration? Direct.

Why is it thought that water is formed in the body? Because less is taken in than is discharged.

What favors its formation? Exercise.

What is the relation of exercise to the temperature? Direct.

Why is it necessary to assume that water is formed in the organism? There is no other way of accounting for the oxygen which is lost in respiration.

In what way is heat produced? By the oxidation of carbon, hydrogen and oxygen.

The heat value of which one is greatest? Hydrogen.

Could the amount of heat that any given quantity of food would produce be estimated? It could.

How? By burning it in oxygen.

How much food should be introduced? Just enough to compensate the waste.

If less is introduced than would keep up the temperature of the body, what change takes place? Body becomes emaciated.

What if more is introduced? Becomes fat.

Is the food oxidized? It is not.

What becomes of the food? It is appropriated.

How does digestion influence temperature? By aiding nutrition.

What is the relation of the temperature of the body to the heat produced? It is lower.

What becomes of the excess of heat? It goes to form force.

In what way is the temperature equalized in the human being? By clothing.

What kinds are worn in summer? Linen.

Why? It is a good conductor for heat.

Kind used in winter? Woolen.

Why? It is a bad conductor and keeps the heat in.

11

If find a patient with sub-normal temperature, what medicine would be given? Belladonna.

Why? It elevates the temperature slightly.

How does it do it? It increases the blood to the heat-gaining area.

What is the heat-losing area? The superficial surface.

What is the heat-producing area? The interior of the body.

In a warm room what is the condition of the blood supply to the heat-losing area? It is increased.

How is it to the heat-producing area? It is lessened.

When going into the cold, what is the first effect? Skin gets red.

What does the heat-producing area do? Gets more blood.

When reaction takes place, how is the skin? Pale.

How does it influence the heat loss? Lessens it.

What does it cause? Congestion of the internal organs.

What name is given to this? Catching cold.

What part is affected? The mucous membrane of the respiratory passages.

Is it always limited to the respiratory passages? No.

What influence has alcohol on temperature? It lowers it.

Why? It carries elements to the tissues to be oxidized instead of the tissue, and it also increases the blood to the heat-losing area.

Why does it cause a feeling of warmth? The blood to the skin is increased.

Through what do we take cognizance of heat and cold? The sensory nerves on the surface.

What kind of sensation is there during a chill? Cold.

How is the temperature? Elevated.

Why do we feel cold? The blood to the surface is lessened.

CHAPTER XXXIII.

NERVOUS SYSTEM.

How is the nervous system divided? *Ans.* Into cerebrospinal and sympathetic.

What is the cerebro-spinal? The brain and cord, and nerves coming from them.

Over what does it preside? The functions of animal life.

What is the sympathetic? The ganglionic system.

Over what does it preside? The functions relating to nutrition.

How is nerve matter divided? Into cells and fibres.

What is the color of a nerve cell? Gray.

Of a nerve fibre? White.

What is the function of a cell? To generate force.

Of a fibre? To conduct force.

Are cells excitable? No.

Give the exceptions? The sympathetic ganglia and certain centers in the brain and cord.

What is necessary for a cell to generate force? It must receive a stimulus.

How are nerve fibres divided? Medullated and nonmedullated fibres.

Name the parts of a medullated fibre? White sheath and substance of Schwann and axis-cylinder.

What is the name given the sheath? White sheath of Schwann.

Give its anatomy? An elastic homogeneous membrane with oval nuclei.

Where·is the sheath absent? In the white part of the brain and cord.

Name of the medullary substance? White substance ot Schwann.

Where is it absent? At the beginning and termination of nerves.

What is the function of the sheath and medulla? They act as an insulator for the cylinder.

What is the axis-cylinder? The central flattened band.

In what nerves is it absent? In the nerves of Remak.

Where is it not surrounded by the sheath and medullary substance? Where the nerves terminate.

How can the axis-cylinder be demonstrated? By applying acetic acid to the cut end.

What is its function? To conduct force.

Where are the nerves of Remak found? In the sympathetic system.

They are connected with what movements? Involuntary.

What is the color of these fibres? Gray.

They resemble what nerves? Embryonic nerves.

Name the anatomical accessory elements of a nerve? Connective tissue, blood-vessels and lymphatics.

Where greatest? Between the muscles.

Why? For protection.

Describe the blood supply to a nerve? The arteries break up into a delicate plexus, which are arranged in longitudinal meshes around the fibres.

Do nerves contain lymphatics? It is supposed so.

What is the function of these accessory elements? To furnish nutrition and afford protection.

Do nerves anastomose? They do.

Where? Near their termination.

What is the peri nerve? Membrane around several fasiculi.

What is the neurilemma? Membrane around a bundle of nerves.

How do nerves terminate in voluntary muscular fibres? Neurilemma and sarcolemma become continuous, the medullary substance stops and the axis-cylinder enters and terminates in an end plate.

How in glands? Form a plexus and pass directly to the glandular cells and terminate in the nucleoli.

How do they terminate in involuntary muscular fibres? Form a plexus in the connective tissue around the muscle and send branches into the nucleoli.

How do sensory nerves end? In integument and in mucous membrane.

Where do you find the corpuscles of Vater? Palms of the hands, soles of the feet and palmar surface of the fingers.

What is their shape? Oval.

What is their relation to the skin? They are beneath the skin.

Describe the termination of nerves in them? The neurilemma surrounds the corpuscle, the medullary substance stops, and the axis-cylinder penetrates the bulb.

What is their function? Have to do with general sensibility.

Where are the tactile corpuscles? Palms of the hands, soles of the feet, nipple, and front of the forearm.

In what layer of the skin? Papillary.

In what kind of papillæ? Compound.

Describe the corpuscles? Oblong, with the long diameter in the direction of the papillæ; the nerve winds around them two or three times and terminates in a pale extremity.

What is their function? Have to do with the sense of touch.

Where are the terminal bulbs? Conjunctiva, buccal cavity, tongue, glans penis and clitoris.

What shape are they? Round.

How do nerves terminate in them? The sheath is continuous with the membrane around the corpuscle, and the axis-cylinder penetrates and terminates in a knotted coil.

What is their function? Have to do with general sensibility.

In what other way do sensory nerves terminate on the skin? In hair follicles and free extremities.

How do they terminate on mucous membranes? They form a plexus beneath the epithelium.

Name the different kinds of nerve cells? Apolar, unipolar, bipolar, and multipolar.

Where do you find the apolar cells? In lower animals.

Where do you find the unipolar and bipolar? In the ganglia of the cranial nerves and posterior roots of spinal nerves.

What is meant by multipolar cells? Those with three or more poles.

Name the kinds? Large and small.

Where find the small? Posterior gray cornua of the cord.

Where find the large? Anterior gray cornua.

With what are the poles continuous? With the axis-cylinder.

How are the poles connected with the cell? By protoplasmic prolongations.

How are the cells connected together? By their poles.

What is the color of a nerve cell? Gray.

Name the accessory elements of a cell? On outer covering, intercellular granular matter, myelocytes, connective tissue, blood vessels and lymphatics.

Give the composition of nerve matter? Cerebrin, protagon, lecithin, xanthine, cholesterin, etc.

If a nerve is cut, is it reproduced? It is.

Describe the new nerve? Like other nerves, except that it has no axis-cylinder.

If nerve cells are destroyed, are they reproduced? No.

What is meant by a motor nerve? One that carries a force to the muscles that causes them to contract.

What is their function ? To give motion to the muscles.

In what direction do they act? From the center to the periphery.

If they are stimulated, what follows ? Motion.

Will motor nerves respond to a stimulus ? They will.

What will paralyze motor nerves ? Curare.

How do they die ? From center to the surface.

If cut, what is the last part to die ? The periphery.

In what time ? Several hours.

What is meant by a sensory nerve ? One that carries a force that causes pain.

If it is irritated, what does it cause ? Pain.

Where is the impression sent? To the brain.

In what direction is it sent ? From center to periphery.

What receives the impression ? The centers in the brain.

What is generated ? Force.

What will paralyze sensory nerves ? Anæsthetics.

In what direction do they act? From periphery to the center.

Why must more chloroform be given in a deep than superficial operation ? Because the deep parts are last affected.

What will paralyze muscular irritability ? Sulpho-cyanide of potash.

How many roots have spinal nerves ? Two.

Which is largest ? The posterior.

What is peculiar about the posterior root ? It has a ganglion.

Where does it unite ? As it leaves the intervertebral foramen.

What does it form ? A mixed nerve.

Which root is motor? Anterior.

Proof? Cut it and will have muscular paralysis.

Which is sensory ? Posterior.

Proof? Stimulate it and will have pain.

If the anterior root is stimulated, why is there slight pain?
There are filaments of communication between the anterior
and posterior root.

Where are the posterior roots largest?   At the ganglion.

What is reflex action?   Action sent back.

Explain it?   The impression is made on the sensory nerves
and is carried to the center, where force is generated and sent
back over the motor nerves.

CHAPTER XXXIV.

CRANIAL NERVES.

What is meant by cranial nerves?   *Ans.* Those passing
through the cranium.

Do they all arise in the cranium?   They do not.

Which does not?   The spinal accessory nerve.

How are they classified?   Into motor and sensory nerves
and nerves of special sense.

What is the other name for the third nerve?   Motor
Oculi communis.

What is its origin?   On the inside of the crura cerebri in
front of the pons Varolii.

With what does it communicate?   The ophthalmic division
of the fifth.

To what is it distributed?   All the muscles of the eye except the superior oblique and external rectus.

What kind of nerve is it at its origin?   Motor.

Before its distribution how is it?   Mixed.

What symptoms are produced by cutting the third nerve?
Phthosis, external strabismus, pupil dilated and directed
down and outward.

Why does the upper lid fall? The levator palpebræ is paralyzed.

Why does the pupil dilate? The circular fibres of the iris are paralyzed.

Does the third nerve supply the iris directly? No.

How does it? Through the lenticular ganglion.

What direction have the fibres of the iris? Circular and radiating.

What nerve supplies the circular? The third nerve.

What supplies the radiating? Sympathetic.

What is the condition of the pupil of the eye if the third nerve is stimulated? Contracted.

If cut, how is it? Dilated.

Why does opium contract the pupil? It stimulates the third nerve.

Why will belladonna cause it to dilate? It stimulates the sympathetic and paralyzes the third.

How is the pupil directed when the third nerve is cut? Downward and outward.

Why? The eye ball is pulled upward and inward by the superior oblique muscle.

What is the other name for the fourth nerve? Patheticus.

What is its origin? Valve of Vieussens near the testes.

Its distribution? To the superior oblique muscle.

With what does it communicate? The ophthalmic nerve.

Is it motor or sensory at its origin? Motor.

How is it at its distribution? Mixed.

What is the origin of the sensory filaments? From the ophthalmic.

What are the symptoms if the fourth nerve is cut? Pupil is directed upward and outward.

Why? The inferior oblique muscle pulls the globe downward and inward.

What is the other name for the sixth nerve? Abducens.

What is its origin? The upper part of the corpus pyramidale.

To what is it distributed? The external rectus muscle.

What is the origin of its sensory filaments? From the ophthalmic.

What are the symptoms if it is cut? Internal strabismus.

Why? The external rectus muscle is paralyzed and the internal pulls the eye in.

What is the other name for the fifth? Tri-facial.

How many roots has it? Two.

Name them? Anterior and posterior.

Which is largest? The posterior.

How many filaments in it? Seventy-five to one hundred.

What does the posterior root do? It forms a ganglion.

What is the origin of the fifth nerve? From the lateral portion of the pons Varolii.

What are the branches of the Gasserian ganglion? Superior and inferior maxillary and ophthalmic.

With what does the inferior maxillary unite? With the motor root.

Where? Just before it gets out of the cranium.

Through what foramen does it pass? Foramen ovale.

What are its branches? External and internal.

What does the external supply? All the muscles of mastication except the buccinator.

What does the internal supply? Mylo-hyoid and anterior belly of the digastric muscle.

What is the nerve of mastication? The motor root of the fifth.

To what muscles is it distributed? The muscles of mastication except the buccinator.

What supplies the buccinator? The facial nerve.

What are the symptoms if the nerve of mastication is cut? Paralysis of the muscles of mastication.

Why will a rabbit die when it is cut? It can't close its mouth.

Why will the teeth grow long? They are not worn off by mastication.

In what way is deglutition impaired by cutting the small root of the fifth? It interferes with the elevation of the os hyoides.

Why? Because the mylo-hyoid and anterior belly of the digastric are paralyzed.

Is the internal branch wholly sensory? It is not.

What muscles do the motor filaments supply? Mylo-hyoid and anterior belly of the digastric.

What is the other name for the facial? Portio dura.

What is its origin? Between the corpus olivary and restiformia.

Where does the facial nerve decussate? In the pons Varolii.

If there is an injury of the brain involving the facial nerve before its decussation, where is the facial paralysis? On the opposite side of the injury.

Where is the hemiplegia? On the same side as the lesion and opposite side of the facial paralysis.

If it is injured after the decussation, where is the facial paralysis? On the same side as the lesion and opposite side of the hemiplegia.

What is the peculiarity of the facial nerve? It has a ganglion.

Where is it? In the hiatus Fallopii.

What is it called? Intumescentia gangliformis.

What is the course of the facial nerve? Toward the ear.

With what does it first communicate? The nervus petrosus superficialis major.

With what next? Nervus petrosis superficialis minor.

What is its third communication? Auricular branch of the pneumogastric.

What does it give to the pneumogastric? Motor filaments.

What does it get from the pneumogastric? Sensory filaments.

What muscle does the tympanic branch of the facial supply? The stapedius muscle.

Where does the chorda tympani go? To the tongue.

What are the terminal branches of the facial nerve? Temporo-facial and cervico-facial.

To what are they distributed? The face from the chin to summit of the head.

What kind of nerve is the facial at its origin? Motor.

Before its distribution? Mixed.

What is the origin of the sensory filaments? From the pneumogastric.

What branch is a nerve of special sense? Chorda tympani.

What nerves have ganglia? Sensory nerves and those of special sense.

What is the portio intermedia? The nerve of Wrisberg.

What is its origin? In the groove between the corpus olivary and restiformia.

With what does it unite? The facial.

What ganglion do they form? Intumescentia gangliformis.

What is the real origin of the chorda tympani? Between the corpus olivary and restiformia.

To what is it distributed? To the tongue.

What is its function? To give taste to the anterior part of the tongue.

What are the symptoms if it is cut? Would have loss of taste at that part.

What gives the taste to the back part of the tongue? The glosso-pharyngeal nerve.

How is mastication affected by cutting the facial nerve? Can not keep the food on the teeth.

Why? Because the buccinator and orbicularis muscles are paralyzed.

How is deglutition affected? The second act is impaired.

Why? The posterior belly of the digastric and the stylo-hyoid muscles are paralyzed.

What part of the second act is impaired? The first part.

What is the function of the terminal branches of the facial nerve? To give motion to the muscles of the face.

What are the symptoms if cut? Paralysis of the muscles of the face.

CHAPTER XXXV.

SPINAL ACCESSORY.

How many roots has it? *Ans.* Two.

What is the origin of the upper? Medulla oblongata.

How does it get into the cranium? Through the foramen magnum.

How does it get out? Through the jugular foramen.

Into what branches does it divide? External and internal.

What does the external supply? Trapezius and sterno-cleido-mastoid muscles.

With what does the internal unite? The pneumogastric.

What kind of a nerve is it at its origin? Motor.

How is it at its distribution? Mixed.

What are the symptoms if the lower root is cut? Paralysis of the sterno-cleido-mastoid and trapezius muscles.

How is respiration affected? It is impeded.

Give the symptoms if the upper root is cut? Loss of voice, heart beats faster and the second act of deglutition interfered with.

How is the voice if one nerve is cut? Hoarse.

How if both are cut? Lost.

What act of deglutition is impaired? The second.

Why? The constrictor muscles of the pharynx are paralyzed.

Why does the heart beat faster? The inhibitory action of the pneumogastric nerve is destroyed.

If the spinal accessory nerve is cut on one side and electricity applied to the pneumogastric, what will be the effect? The heart stops beating.

How long before the heart is influenced? Nine or ten days.

CHAPTER XXXVI.

HYPOGLOSSAL.

What is its origin? *Ans.* Between the corpus olivary and pyramidale.

How does it get out of the cranium? Through the anterior condyloid foramen.

What are its branches? Descendens noni, thyro-hyoid and muscular.

What organ is supplied by it? The tongue.

What kind of nerve is it at its origin? Motor.

How at its distribution? Mixed.

What are the symptoms if it is cut? Paralysis of the tongue.

How is mastication influenced? Can't pull the jaw down.

Why? The depressors of the os-hyoids are paralyzed.

CHAPTER XXXVII.

TRI-FACIAL.

What kind of nerve is the posterior root? *Ans.* Sensory.

To what is it distributed? To the face.

What are the symptoms if it is cut posterior to the ganglion? Loss of sensation to the parts that it is distributed.

What symptoms if cut the ganglion or anterior to it? Loss of sensation and purulent inflammation of the eyes.

Why? Because you have bad nutrition together with increased blood supply to the face.

Why is nutrition bad? Can't masticate.

Why can't one masticate? In cutting the posterior root the anterior is always cut and the muscles of mastication are paralyzed.

Why are the blood-vessels dilated? The vaso-motor constrictor nerves to the blood-vessels of the face pass up from the neck through the ganglion, and when they are cut the blood supply is increased.

If irritation is applied to the soft palate, what act is excited? Deglutition.

How is it brought about? The impression is made on the tri-facial, one of the reflex nerves of deglutition which carries it to the center, where force is generated which causes efforts at deglutition.

What effect has cutting of the tri-facial? One of the reflex nerves of deglutition is paralyzed, which interferes with deglutition.

How many reflex nerves of deglutition are there? Two.

Name them? Superior laryngeal and tri-facial.

Why was the fifth nerve called tri-facial? Because it makes its appearance at three places on the face.

CHAPTER XXXVIII.

PNEUMOGASTRIC.

What is its origin? *Ans.* Between the corpus restiformia and olivary.

How does it get out of the cranium? Through the jugular foramen.

What kind of nerve is it at its origin? Sensory.

How is it at its distribution? Mixed.

What is the source of the motor filaments? Spinal accessory and facial.

Give its communications? Spinal accessory, facial and sympathetic.

What is its first branch? Auricular.

What are the communications of the auricular? Facial and glosso-pharyngeal.

To what is it distributed? External auditory meatus and membrana tympani.

What is it function? To give sensation to the parts.

What is the second branch of the pneumogastric? Pharyngeal.

What plexus does it help to form? The pharyngeal plexus.

What forms the pharyngeal plexus? The pharyngeal branches of the pneumogastric and glosso-pharyngeal and the symphathetic.

What does the plexus supply? The muscles of the pharynx.

The motor filaments of the plexus comes from which branch? Pharyngeal branch of the pneumogastric.

What is the real source? The spinal accessory.

Which root of the spinal accessory? Upper.

What are the symptoms if the pharyngeal branch of the pneumogastric is cut? The muscles of the pharynx are paralyzed.

What if the upper root of the spinal accessory is cut? Same symptoms.

What is the third branch of the pneumogastric? Superior laryngeal.

To what organ is it distributed? Larynx at the upper part.

What is its function? To give sensation to that part so as to protect the larynx during deglutition.

During what act of deglutition? The second act.

What are the symptoms if it is cut? Loss of sensation to the mucous membrane of the larynx and one of the reflex nerves of deglutition is destroyed.

What if stimulated? Will make efforts at deglutition and cough.

By cutting it, how is deglutition influenced? It is interfered with, as one of the reflex nerves of deglutition is cut.

What branch does the superior laryngeal nerve give off? The external laryngeal.

To what is it distributed? The inferior constrictor of the pharynx and crico-thyroid muscles and the thyroid gland.

With what does it communicate? Inferior laryngeal and sympathetic.

What is the function of the external laryngeal branch? To animate the crico-thyroid and the inferior constrictor of the pharynx.

What is the fourth branch of the pneumogastric? Inferior laryngeal.

What is its other name? Recurrent laryngeal.

What is its course on the left side? Around the arch of the aorta.

12

What is it on the right side? Around the sub-clavian artery.

To what is it distributed? All the muscles of the larynx except the crico-thyroid muscle.

With what does it communicate? With the external laryngeal and the sympathetic nerves.

Result if cut in young animals? Death.

Why? The cartilages are soft and collapse.

What if cut in adult animals? They border on suffocation.

What is the nerve of voice? The recurrent laryngeal.

Where do the filaments come from that cause voice? From the spinal accessory.

Proof? Cutting the spinal accessory the inferior laryngeal produces the same effect.

Does the recurrent laryngeal contain any other motor filaments? It does.

Proof? Because if the spinal accessory is cut above, the symptoms are not as bad as cutting the inferior laryngeal.

Why does the external laryngeal arise from the superior? It goes down to the crico-thyroid to prepare the larynx for speaking.

Why does the recurrent laryngeal go down and wind around the aorta on the left and sub-clavian artery on the right side? To transmit its force from below, up.

During what act of respiration do we speak? Expiration.

Where does the force come from? From below.

What is the depressor nerve of the circulation? Nerve of Cyon and Ludwig.

What is its origin? From the superior laryngeal and pneumogastric.

To what is it distributed? To the heart.

What effect has cutting it on the heart's action? None.

What if electricity is applied to the peripheral cut end? No effect.

To central end? It reduces the pulsation.

What effect has cutting of the nerve on arterial pressure? None.

If the central end is stimulated, what is the result? Reduces arterial pressure.

What effect if the depressor nerve is stimulated after the great splanchnic has been cut? It is diminished, but not as much as when the splanchnic is intact.

What is the fifth branch of the pneumogastric? Cardiac branches.

What do they form? Cardiac plexus.

They supply what organ? The heart.

How is the heart affected by cutting the cardiac branches of the pneumogastric? It beats faster.

What will it do if the pneumogastric is cut? It beats faster.

What if upper root of spinal accessory? It beats faster.

From what nerve does the inhibitory action of the pneumogastric come? The spinal accessory.

What is the sixth branch of the pneumogastric? The pulmonary branches.

What forms the anterior pulmonary plexus? The anterior pulmonary branches of the pneumogastric and branches from the cardiac plexus.

What forms the posterior pulmonary plexus? Posterior pulmonary branches, and branches from the third and fourth thoracic ganglia.

What is the effect of cutting the pulmonary branches? The lung becomes carnified.

Name the seventh branch of the pneumogastric? Œsophageal.

What are the symptoms if it is cut? Paralysis of the lower part of the œsophagus.

What are the last branches of the pneumogastric? Gastric.

What are the symptoms about the stomach if the pneumogastric is cut? Secretion is arrested and the walls are paralyzed.

Why are the walls paralyzed? Because the motor nerve of the stomach gets there through the pneumogastric.

Is the pneumogastric nerve the motor nerve to the stomach? No.

What is the source of the motor nerve of the stomach? From the sympathetic.

How does it get there? Through the pneumogastric.

What is the inhibitory nerve to the blood-vessels of the stomach? The pneumogastric.

What effect has cutting it? The blood vessels contract.

What effect if it is stimulated? The blood vessels dilate.

What if the sympathetic is cut? Dilates them.

What if the sympathetic is stimulated? They contract.

What are the symptoms in the bowels if the pneumogastric is cut? Loss of motion

How are the blood-vessels influenced? They contract.

How are they influenced if it be stimulated? They dilate.

If the pneumogastric is cut between the lungs and liver, how is the glycogenic function of the liver influenced? It is not affected.

What if the pneumogastric is cut in the neck? Its function is arrested.

What is the effect of injury of the floor of the fourth ventricle of the brain? Causes diabetes mellitus.

In clap, why is there pain in the rectum? Branches of the internal pudic nerve supply both parts.

Why on the tuber ischii? The inferior pudendal nerve is affected.

After operation for piles, why is there pain in the head of the penis? Both parts are supplied by the internal pudic nerve.

Where is the pain in cystitis? At the neck of the bladder.

In irritation of the neck of the bladder, where is the pain? In the head of the penis.

Why? Injury of the sacral nerves, which supply both parts.

When a nerve is injured, where is the pain expressed? At its distribution.

In hip-joint disease, why is there pain at the knee? The obturator nerve supplies both parts.

In pleuritis, why have pain on the inside of the arm? The intercosto humeral nerve communicates with the nerve of Wrisberg.

In inflammation of the diaphragm, where is the pain? Over the clavicle, sternum and acromium process.

Why? Due to communication of the phrenic nerve with the cervical plexus.

If there is pain on the back part of the ear or immediately in back of it, where is the injury? In the neck, involving the cervical plexus.

Give the rationale of a blister and plaster? They relieve tension by dilating the capillaries, giving better blood supply.

Why is the glutens maximus muscle supplied by the lesser ischiatic nerve? So that in sexual intercourse that muscle will act in unison at the crisis or termination of the act of copulation.

Why does the spinal accessory nerve go through the cranium? To become the nerve of voice.

In disease of the liver, where is the pain? At the lower part of the scapula.

Why? The great splanchnic nerve is involved, supplying both parts.

CHAPTER XXXIX.

SPINAL CORD.

What is the cerebro-spinal axis? *Ans.* The nervous matter in the cranium and spinal canal exclusive of the cranial and spinal nerves.

What membranes surround it? Dura mater, arachnoid and pia-mater.

What is the relation of the fluid to the axis? It surrounds it.

Where is the gray matter in the brain? On the outside.

Where is it in the cord? On the inside.

What kind of membrane is the dura mater? Fibro serous.

How many fibro-serous membranes in the body? Three.

Name them? Pericardium, dura mater and tunica albuginea.

Give its physiological anatomy? Dense fibrous inelastic membrane.

Where is it attached to the bone? In the cranial cavity and at the foramen magnum.

Give its peculiarity in the spinal canal? Hangs loosely around the cord.

Describe the arachnoid? A delicate serous membrane with the surfaces almost in contact.

Where is the fluid? Beneath both layers.

Name the fluid? Cephalo-rachidian fluid.

Describe the pia-mater? A delicate fibrous membrane which is very vascular.

What is its function? To give nutrition to the brain and cord.

How does it differ around the cord? Thicker, stronger, less vascular and more closely adherent.

What is it somtimes called? The neurilemma of the cord.

What forms the sheath of spinal nerves? The pia-mater.

What is its arrangement around the cord? It is in bands.

What is the ligamentum denticulatum? A narrow band between the anterior and posterior roots of the spinal nerves extending from the foramen magnum to the filum terminale.

Is the quantity of blood in the brain subject to variation in health and disease? Yes.

During what act of respiration is the volume of the encephalon greatest? Expiration.

Why? It favors arterial flow and retards venous.

When does it decrease? During inspiration.

Why? Inspiration increases venous and retards arterial circulation.

Give the peculiarity of the encephalic capillaries? They are surrounded by a perivascular system and lymphatics.

Why? To equalize the pressure.

Name the fluid of the brain and cord? Cephalo-rachidian fluid.

Where is the greatest amount situated? In the subarachnoid space.

How do the ventricles of the brain communicate with the central canal? At the floor of the fourth ventricle.

How much of the fluid is there? About two fluid ounces.

What are the symptoms if it be drawn out forcibly with a syringe? The movements are at first impeded and finally have paralysis.

Why? You have a congestion of blood-vessels, followed by effusion.

What if there is a sudden increase of the fluid? Coma.

Why? It compresses the centers.

Give the properties of the cephalo - rachidian fluid? Transparent, colorless, non-viscid alkaline fluid of a saline taste, and it resists putrefaction for a long time.

What is its function? To equalize blood pressure.

Where do you find the spinal cord? In the spinal canal.

How long is it? Fifteen to eighteen inches.

How much does it weigh? One and one-half ounces.

What is its extent? From the foramen magnum to the first lumbar vertebra.

Name its enlargements? Medulla oblongata and the brachial and lumbar enlargements.

What is the filum terminale? The slender gray termination.

What is the chorda equina? The sacral and coccygeal nerves which pass downward from their origin.

Where is the white matter of the cord? Outside.

Where is the gray? Inside.

Where is the proportion of the white to the gray greatest? Above.

Why? The fibres going to the cord increase it from below up.

Where is the cord wholly gray? At its inferior extremity.

Name the fissures? Anterior and posterior median fissures.

What is the extent of the anterior? One-third the thickness of the cord.

What does it receive? A fold of pia-mater.

What is the extent of the posterior? One-half the thickness of the cord.

What fills it? Connective tissue and blood-vessels.

How is the cord divided? Into two lateral halves.

How is each half divided? Into anterior, posterior and lateral columns.

Give the boundaries of the anterior column? Anterior median fissure and the motor root of the nerves.

What bounds the lateral columns? The anterior and posterior roots of the spinal nerves.

Give them of the posterior column? Posterior median fissure and the posterior root of the nerves.

What shape is the gray matter of the cord? Like the letter H.

Describe the anterior cornua? Broad and does not come ·near the surface.

Describe the posterior? Narrow and long, and comes near the surface.

What connects the gray matter? Gray commissure.

How many gray commissures are there? Two.

What is in the center? Central canal.

What lines the canal? Epithelium.

It communicates above with what? Fourth ventricle of the brain.

What surrounds the central canal? Connective tissue.

What is the composition of the white matter? Made up of nerve fibres.

What cells are found in the anterior gray cornua? Large multipolar.

Are they motor or sensory? Motor.

What cells are in the posterior cornua? Small multipolar cells.

What is the function of the cord? To act as a center and a means of communication between the brain and the extremities.

Through what part are impressions carried? Gray matter and posterior roots.

The force is sent back to the muscles through what part? The anterior and lateral columns.

What other function has the cord? To act as a center.

What part acts as a center? The gray matter.

How does the anterior root of the nerve arise from the large multipolar cells? The axis-cylinder is continuous with the poles.

How are the cells connected with the brain? Nerve filaments are sent up.

Through what part do they pass? Anterior and lateral columns.

Does the posterior root of the nerve arise from the cell? No.

From what does it arise? From number of nerve filaments coming from the cell.

In what way are the small multipolar cells connected with the brain? Nerve filaments are sent up.

Through what part do they pass? Gray matter.

What parts of the cord are excitable? Anterior and lateral white columns.

What is meant by excitable? When stimulated it causes motion.

Is this part sensitive? No.

What part is sensitive? Posterior column.

Where is it most sensitive? Near the root of the nerve.

Is this part excitable? No.

Is the gray matter excitable and sensible? No.

What is the function of the anterior and lateral columns? To transmit motor nerves.

Proof? Cut them and it will cause loss of motion below.

What are the symptoms if the lateral columns in the cervical region are cut? Paralysis of motion on the side cut.

If the anterior columns are cut? Only diminished motion.

If cut lateral column in the dorsal region? Diminished motion on the same side.

If cut anterior column in the dorsal region? Loss of motion below.

What is hemiplegia? Paralysis of one side of the body.

The fibres from the right side of the brain pass to which side of the cord? To the left side.

Where do they cross over? In the medulla oblongata.

What is paraplegia? Paralysis of the body from the thigh down.

What is the function of the gray matter of the cord? To act as a center.

Proof? Sever the head from the body, you still have motion.

Where do the sensory filaments decussate? In the gray matter all up and down the cord.

Proof? Split the cord from the anterior to posterior median fissure, it causes loss of sensation on both sides.

What is the function of the posterior columns of the cord? To preside over muscular co-ordination.

Proof? Cutting it causes locomotor ataxia.

What is tabes dorsalis? Loss of muscular co-ordination.

What are the symptoms of cutting half the cord in the cervical region? Loss of motion and hyperæsthesia on the same side and loss of sensation on the opposite side.

If cut in the dorsal region? Same as in the cervical.

Where is the loss of motion? On the same side.

The loss of sensation? On the opposite side.

Where is the hyperæsthesia? On the same side that is cut.

How is sensation affected by cutting the gray matter from the medulla to the lower portion in a vertical direction? It is lost on both sides.

How is motion affected? It is not affected.

How is motion affected by cutting vertically between the two anterior pyramids of the medulla oblongata? It is lost on both sides below.

How is sensation affected? It is not affected.

What effect has decapitation on sensation? Lost.

Why? The brain only appreciates pain.

What effect has it on voluntary motion? It is lost.

Why? The brain is absent.

Does it destroy the function of the motor and sensory nerves? No.

What part of the nervous system generates force? The cells.

Does severing the head from the body destroy motion? No.

What is this motion called? Reflex motion.

Are reflex movements confined to the cord? No.

Give illustration? The movements of the iris are reflex and are confined to the brain.

What kind of movement do we mean when we say reflex? Its cause originates on the surface and it is involuntary.

What conditions are necessary for movements to take place in a frog after decapitation? An irritant must be applied to the surface.

What condition is necessary for reflex movements to take place from the cord? The roots of one or more of the nerves must be intact.

Under what conditions will movements take place when all the nerves have been severed but one? There must be a great stimulus.

What effect has decapitation on reflex excitability? It increases it.

How is the lower part of the cord affected by section in the dorsal region? Its reflex action is increased.

What effect have opium and strychnine poisoning? They increase it.

What effect have anæsthetics? They diminish it.

How can you tell complete anæsthesia? The reflex of the eye is lost.

What is the effect of irritating the soles of the feet in paraplegia? It causes motion of the legs.

On what do the movements of deglutition depend? On reflex action.

On what does the ejaculation of semen depend? Reflex action also.

Give an illustration of reflex action of the sympathetic system? The action of the bowels or defæcation.

CHAPTER XL.

BRAIN.

What is the function of the gray matter? *Ans.* To generate force.

How are the impressions carried to the brain? Through the gray matter in the cord and the cranial nerves.

Does the brain have to do with reflex action? Yes.

Illustrate? Contraction of the iris is reflex.

What is the weight of the brain in the male? Forty-eight to fifty-two ounces.

How much in the female? Forty-three to forty-seven ounces.

How much did the largest brain weigh? Seventy-two ounces.

Give its weight at birth? Ten to eleven ounces.

Give the rapidity of development at different periods? It increases to forty, then it decreases.

What is the proportion of the size of the cerebrum compared to the cerebellum in the male? It is eight and four-seventh times larger.

How in female? Eight and one-fourth times larger.

What is the appearance of the surface of the cerebrum? It is convoluted.

Where is the gray matter? On the outside.

How thick is it in a convolution? One-twelfth to one-eighth inch.

Where are the sensory cells? On the surface.

Where are the motor cells? Deeper.

Is the brain excitable? No.

Is it sensitive? No.

Exception? Some of the ganglia.

What is the function of the cerebrum? To preside over intellect.

What effect has extirpation on hearing, sight and smell? They are slightly blunted.

Effect on sensation and motion? None.

If the brain is destroyed, is it reproduced? No.

What are the symptoms of hemorrhage in the corpus striatum? Have loss of motion on the other side.

Does the brain weigh more in man or inferior animals? In man.

Exception? The brain of the elephant and the whale.

What is the proportion of the size of the cerebrum in man and inferior animals? It is larger in man.

What appearance has the surface of the brain in inferior animals? It is smooth.

What is the relation of the number and depth of the convolutions to the intellect? Direct.

The development of the cerebrum in different races of men is in what proportion? To the degree of civilization.

What is the relation of the size of the brain to the intellect? Usually direct.

Is the rule absolute? No.

Give an exception? The largest brain on record was that of an idiot.

What has exercise to do with the development of the brain? The development depends on exercise.

What faculty has been located in the brain? The power to articulate language.

Where is it located? In the third convolution of the anterior lobe of the left hemisphere.

Proof? If destroyed, speech is usually lost.

Is its location definite? No.

Why? Because when the supposed location is injured you sometimes have speech, and injury to other parts sometimes causes the loss of speech.

CHAPTER XLI.

CEREBELLUM.

What is its color on the surface? *Ans.* Gray.

Internally? White.

Describe the superior peduncle? It connects the cerebellum with the cerebrum.

Describe the middle peduncle? It connects the two hemispheres of the cerebellum.

Describe the inferior? It connects the cerebellum with the corpora restiformia.

How are the sides of the cerebellum connected? By the middle peduncle.

Give the properties of the cerebellum? Neither excitable nor sensible.

What is its function? To preside over muscular-co-ordination and reproduction.

What is the function of the corpora striata? They have to do with motion.

What is the function of the thalami optici? They have to do with sensation.

Function of the corpora quadrigemina? Has to do with sight.

Give the properties of the pons Varolii? Excitable and sensible on its posterior part.

What is its other name? Tuber annulare.

What is its general function? Has to do with motion and sensation.

Does it act as a nerve center? Yes.

Proof? If the cerebrum is removed, can still have motion.

CHAPTER XLII.
MEDULLA OBLONGATA.

What is it? *Ans.* The upper dilated portion of the spinal cord.

How long is it? One and one-half inches.

How wide? Three-fourths to one inch.

What is its position? On the basilar process of the occipital bone.

How is it divided? Into two halves.

Name the fissures? Anterior and posterior median fissures.

With what are the anterior pyramids continuous? The lateral and anterior white columns of the cord.

Where do the nerves decussate? In the median line.

Where are the corpora olivaria? External to the corpora pyramidale.

What shape are they? Oblong.

What surrounds them? A groove.

What color are they on the surface? White.

On the interior? Gray.

Where are the restiform bodies? External to the olivary bodies.

With what are they continuous? Posterior white column of the cord.

Where is the lateral tract? Beneath the olivary body between the corpus pyramidale and restiformia.

Where is the respiratory center? In the lateral tract.

What is the function of the medulla? Has to do with reflex action, respiration, heart's action and deglutition.

CHAPTER XLIII.

SPECIAL SENSE.

Give the origin of the olfactory nerve? *Ans.* By three roots, one from the carunculæ mammillares in the anterior lobe, the inner from the anterior lobe near the substantia perforata, and the external from the anterior commissure in the third ventricle.

Give its distribution? The mucous membrane of the nose.

Give its properties? It is a nerve of special sense.

Its function? To give to the nose the sense of smell.

Name the nerves of taste? Chorda tympani and glossopharyngeal.

What is the origin of the chorda tympani? From the facial.

To what part of the tongue is it distributed? Anterior part.

What is the origin of the glosso-pharyngeal? Between the corpus olivary and restiformia.

What are its properties at its origin? Special sense.

13

At its distribution ? Mixed.

What is its function ? Give taste to the back part of the tongue.

Give the mechanism of taste? Food comes in contact with the terminal filaments of the nerves of taste, which carry the impression to the brain.

What is the origin of the optic nerve ? Corpus genicula-tum externum, thalamus opticus, brachium anterius and nates.

Distribution? To the retina of the eye.

What kind of nerve is it? Nerve of special sense.

What is its function? To give to the eye the sense of sight.

CHAPTER XLIV.

SYMPATHETIC.

What does it mean? Suffering for another.

What other names has this system? Ganglionic or vege-tative system.

What is its arrangement? In a double chain, up and down the vertebral column.

To what is it distributed? Non-striated muscular fibres.

Give the exception? It is distributed to the heart.

What is the color of a sympathetic nerve? Gray or red.

What is peculiar about its fibres? They are smaller than other fibres.

What is peculiar of the cells? They are mostly bipolar.

Are they motor and sensory? They are motor.

Proof of each? Can have motion or pain by stimulating them.

How does the motion differ from that of the cerebro-spinal system? It comes on slower and is less powerful.

What is the origin of the action of the sympathetic? From the cerebro-spinal.

What regulates the pupil of the eye? The cerebro-spinal and sympathetic nerves.

All the motor fibres come from what ganglion? Ophthalmic.

What is meant when the term vaso-motor is used? Motion to the vessel.

What are the symptoms if the superior cervical ganglion is destroyed? An increased blood supply to that side of the head.

What is the origin of part of its power? From the cervical ganglia below.

What is the origin of the vaso-motor fibres of the neck? The spinal cord.

How is it connected with the sympathetic? Each ganglion sends communicating branches to the spinal nerves.

What are the symptoms if the first and second dorsal nerves are cut in the spinal canal? The blood vessels on that side become dilated.

What if the trunk of the sympathetic is cut between the second and third ribs? They become still more dilated.

From where do the sensory and motor nerves of the head and face come? Cranial nerves.

Where do the nerves to the blood-vessels come from? From the spinal cord below.

Proof? If the second and third dorsal nerves are cut before they get out of the spinal canal, the blood-vessels of the head and face become widely dilated.

What is the origin of the motor and sensory nerves of the arm? From the brachial plexus.

What is the origin of the nerves to the blood vessels of this part? From the cord as low down as the fourth and fifth dorsal nerves.

Proof? If the nerves forming the brachial plexus are cut before they get out of the spinal canal, there is loss of motion and sensation in the arm, but if the cord is injured as low as the fourth and fifth dorsal nerves, the blood supply to the arm is increased.

What is the origin of the motor and sensory nerves of the lower leg? From the sacral plexus.

What is the origin of the nerves of the blood-vessels to this part? From the cord higher up.

Proof? By cutting the nerves forming the sacral plexus there is loss of motion and sensation in the leg, while cutting the nerves higher up the blood supply to the part is increased.

What are the symptoms if the vaso-motor or sympathetic nerves to the blood-vessels of the salivary glands are cut? The vessels dilate.

Effect of cutting the sympathetic in the neck? The blood-vessels of the neck become dilated.

Effect of cutting the sympathetic nerves going to the heart? It beats slower.

If cut those going to the stomach? It increases digestion.

If cut the great splanchnic nerve? The blood vessels of the guts become dilated.

What are the nerves that cause dilatation of the blood-vessels? Vaso-motor dilators.

What is their origin? From the cerebro-spinal system.

What is the dilator nerve of the blood-vessels of the salivary glands? The chorda tympani nerve.

Of the heart? The pneumogastric nerve.

Of the blood vessels of the stomach? The pneumogastric.

What are the symptoms if it is cut? Have diminished blood supply and loss of motion.

Nerves from what systems are distributed to blood-vessels of the stomach? Cerebro-spinal and sympathetic.

What are the symptoms if the cerebro-spinal is cut? The blood-vessels contract.

What if it is stimulated? They dilate.

What are the symptoms if the sympathetic is cut? The blood-vessels dilate.

What if the dilator or inhibitory nerve is cut? They contract.

If it is stimulated, what effect? They dilate.

What does the inhibitory counteract when stimulated? The sympathetic.

What causes the vessels to dilate? The pressure of the blood in the arteries.

Explain the action of the nerves on sphincter muscles? When the sympathetic is stimulated it narrows the opening, as it causes contraction of the circular fibres.

Explain the reflex dilatation of blood-vessels? When food is taken in the mouth the sensory nerves receive the impression and carry it to the center, which sends out force to the blood-vessels of the salivary glands and causes them to dilate.

CHAPTER XLV.

REPRODUCTION.

What is reproduction? *Ans.* Reproduction is the function by which the species is preserved, and is accomplished by the organs of generation in the two sexes.

As regards physiology, what can you say of the so-called spontaneous generation? Formerly this question was much discussed, especially in relation to some of the lower animals; this, however, is now of purely historical interest. As actual knowledge of facts have accumulated, the limits of what was thought to be spontaneous generation have become more and more restricted, until now it is generally admitted that it does not exist in the history of animals.

Give a certain distinct form of generation? There are other certain distinct forms of generation, but the only one that has any considerable importance in connection with human physiology is generation of new beings by the union of male and female elements in the fecundation of the ovum, with the development of the fecundated ovum.

This is known as what? Known as sexual generation. The two elements of generation are developed in separate beings, male and female, and those elements are brought together normally, in what is known as sexual connection or copulation.

CHAPTER XLVI.

FEMALE ORGANS OF GENERATION.

What is most important for a comprehension of those special organs of generation? *Ans.* A knowledge of certain points in the anatomy is essential to the comprehension of the most important of the process of generation.

As regards the functions of generation, to what extent does nature depend upon the male, following a fruitful intercourse of the sexes? Following a fruitful intercourse, as regards the male, ceases with the comparatively simple process of penetration of the male element through the protective covering of the ovum and its fusion with the female element.

After the two element come in contact, what follows? The fecundated ovum then passes through certain changes, which are the first process of its development, forms its attachments to the body of its mother, continues its development and is nourished and grows until the fœtus at term is brought into the world.

How are the female organs of generation divided? Anatomically, into external and internal ; thus, the external organs are the vulva, adjacent parts, and the vagina. The internal are the uterus, the Fallopian tubes and the ovaries.

What is most important? The ovaries are the true female organs of generation, in which alone the female element can be produced, although the Fallopian tubes and the uterus are accessory in their use, the ovum passing through the Fallopian tubes to the uterus, where it forms attachments to the body of the mother, which are essential to its nourishment and full development after fecundation.

Please give the anatomy in detail of the female organs of generation? The ovaries are two small ovoid, flattened bodies, measuring one inch and a half in length by three-

quarters of an inch in width. They are situated in the cavity of the pelvis and imbedded in the posterior layer of the
brood ligaments, and attached to the uterus by a round ligament. The ovary consist of an external membrane of fibrous
tissue, the *cortical* portion, in which are imbedded the
Graafian vesicles and the internal portion. The *stroma* contains blood-vessels. The weight of each ovary is from sixty
to one hundred grains (3-9 to 6-5 grammes), and these organs
are largest in the adult virgin. Its attached border is called
hilum, and at this portion the vessels and nerves penetrate,
and its surface is marked by rounded, translucent elevations,
produced by distended Graafian follicles, with little cicatrices
indicating the situation of ruptured follicles.

GRAAFIAN FOLLICLES.—These vesicles or follicles were
described and figured by Dr. Graafe in 1672, and are known
by his name. They are exceedingly numerous, but situated
only in the cortical portion. Although the ovary contains
the vesicles from the period of birth, it is only at the period
of puberty that they attain their full development. From
this time onward to the catamenial period there is a constant growth and maturation of the Graafian follicles.
They consist of an external investment, composed of fibrous
tissue and blood-vessels, in the interior of which is a layer
of cells forming the *membrana granulosa*. At its lower portion there is an accumulation of cells, the *proligerous disc*,
in which the *ovum* is contained. The cavity of the vesicle
contains a slightly yellowish alkaline, albumnious fluid. In
the process which culminates in the discharge of the ovum
into the fimbriated extremity of the Fallopian tube, the
Graafian follicle gradually enlarges, becomes distended with
liquid and finally breaks through and ruptures upon the surface of the ovary.

THE PAROVARIUM.—The Parovarium or organ of Rosen-
muller is simply the remains of the Wolffian body, lying in

the folds of the broad ligaments, between the ovary and the Fallopian tube. It consists of twelve to fifteen tubes of fibrous tissue, lined by ciliated epithelium. It has no physiological importance.

THE UTERUS.—The uterus is a pear-shaped body, somewhat flattened antero-posteriorly, presenting a fundus, a body and a neck.

At its lowest extremity is an opening into the vagina, called the os-externum. At the upper portion of the neck is a constriction which indicates the situation of the os-internum. The uterus is usually about three or three and a half inches long, two inches in breadth at its widest portion, and one inch in thickness. Its weight is from one and a half to two and a half ounces (42·5 to 71 grammes). It is somewhat loosely held in place by the broad and round ligaments and by the folds of the peritoneum in front and behind the delicate layer of peritoneum which from its outer covering extends behind as far down as the vagina, where it is reflected back upon the rectum, and anteriorly, a little below the upper extremity of the neck (os internum), where it is reflected upon the urinary bladder.

The muscular walls of the uterus are composed of nonstriated fibrous tissue arranged in several layers. They are so closely bound together that they are isolated with great difficulty, and the muscular tissue of the uterus is remarkable from the fact that the fibres enlarge immensely during gestation, becoming at that time ten or fifteen times as long, and five to seven times as broad as they are in the unimpregnated state.

The mucous membrane of the uterus is of a pale, reddish color, and that portion lining the body is smooth, and is so closely attached to the subjacent structures that it can not be separated to any great extent by dissection.

The changes which the mucous membrane of the uterus

undergoes during menstruation are remarkable. Under ordinary circumstances its thickness is one-twenty-fifth to one-
fourteenth of an inch, but it measures during the menstrual
period one-sixth to one-fourth of an inch.

The blood-vessels of the uterus are very large and present
certain important peculiarities in their arrangement. The
uterine arteries pass between the layers of the broad ligament
to the neck, and then ascend by the sides of the uterus, presenting a rich plexus of vessels, anastomosing above with
branches from the ovarian arteries, sending branches over
the body of the uterus, and finally penetrating the organ, to
be distributed mainly in the middle layer of muscular fibres.
In their course these vessels present a convoluted arrangement and form a sort of mold of the body of uterus.

In the muscular walls of the uterus are large veins, the
walls of which are closely adherent to the uterine tissue.
During gestation these vessels become enlarged, forming the
so-called uterine sinuses.

Lymphatics are not very abundant in the unimpregnated
uterus, but they become largely developed during gestation.

The uterine nerves are derived from the inferior hypogastric and the spermatic plexus, also the third and fourth sacral. In the substance of the uterus they present in their
course small collections of ganglionic cells, and it is said that
the nerves pass finally to the nucleoli of the muscular fibres.

THE FALLOPIAN TUBES—Are about four inches in length
and extend outward from the upper angles of the uterus, between the folds of the broad ligaments, and terminate in a
fringed extremity, which is attached by one of the fringes to
the ovary. They consist of three coats, (1) the external or
peritoneal, (2) middle or muscular, the fibres of which are
arranged in a circular or longitudinal direction, (3) internal
or mucous, covered with ciliated epithelial cells, which are
always waving in a kind of antiperistaltic action from the

ovary to the uterus, and opening into the uterus upon either side at the cornua. They present a small orifice about one-twenty-fifth of an inch in diameter. From the cornua they take a somewhat undulatory course outward, gradually increasing in size, so that they are rather trumpet-shaped.

The extremity next to the ovary is marked by ten or fifteen fimbriæ or fringes, which have given this portion the name of the fimbriated extremity or *morsus diaboli*. All of these fringe-like processes are free except one, and this one, which is longer than the others, is attached to the outer angle of the ovary, and presents a little gutter or furrow, extending from the ovary to the opening of the tube. At this extremity is the abdominal opening of the tube, which is two or three times as large as the uterine opening. Passing from the uterus, the caliber of the tube gradually increases, as the tube itself enlarges, and there is an abrupt constriction at the abdominal opening.

EXTERNAL ORGANS OF THE FEMALE.—It is not necessary to give a minute description of the female external organs of generation. The opening by the vulva externally, and terminating at the neck of the uterus, is a membraneous tube, the *vagina*. This lies between the bladder and the rectum. It has a curved direction, being three and one-sixth to three and one-half inches (eight to nine centimeters) long in front, and three and one-half to four inches (nine to ten centimeters) long posteriorly (Sappey). At the constricted portion of the outer opening there is a muscle called the sphincter vaginæ, and the tube is somewhat narrowed at its upper end, where it embraces the cervix uteri. The inner surface presents a mucous membrane marked by transverse rugæ, with papillæ and mucous glands. Its surface is covered with flattened epithelium. The vagina is quite extensible, as it must be during parturition, to allow the passage of the child.

The parts composing the external organs are abundantly supplied with vessels and nerves. In the clitoris, which corresponds to the penis in the male, and on either side of the vestibule, there is a true erectile tissue.

The Ovum.—The ovum lies in the Graafian follicles imbedded in the mass of granular cells which form the discus proligerous. Surrounding the ovum are cells similar to those found in the other parts of the membrana granulosa and two or three layers of columnar cells, the latter lying next the zona pellucida. These columnar cells constitute the corona radiata (Bischoff).

The ovum itself presents the following structures : (1) zona pellucida, (2) perivitelline space, (3) a clear outer zona of the vitellus, (4) protoplasmic zone (formative yelk), (5) dentoplasmic zona (nutritive yelk), (6) germinal vesicle or germinal spot. The extremely thin membrane within the zona pellucida and immediately surrounding the vitellus, described under the name of vitelline membrane by some anatomists, was not observed by Nagel in the human ovum. The ovum is globular, with a diameter of about one one-hundred and fiftieth of an inch, measured from the outer border of the zona pellucida.

As the Graafian vesicle matures, it increases in size, forms an augmentation of its liquid contents and approaches the surface of the ovary, where it forms a projection measuring from one-fourth to one-half an inch in size. The maturation of the vesicle occurs periodically, about every twenty-eight days, and is attended by the phenomena of menstruation. During this period of active congestion of the reproductive organs, the Graafian vesicles rupture, the ovum and liquid contents escape and are caught by the fimbriated extremity of the Fallopian tubes, which has adapted itself to the posterior surface of the ovary. The passage of the ovum through the Fallopian tube into the uterus occupies from ten to fourteen days, and is accomplished by the muscular contraction and the action of the ciliated epithelium.

The fact that the ovum in the great majority of instances passes into the Fallopian tubes is sufficiently evident. The fact, also, that ova may fall into the cavity of the peritoneum is illustrated by occasional occurrences of extrauterine pregnancy, a rare accident, which shows that in all probability the failure of unimpregnated ova to enter the tubes is exceptional. As regards the mechanism of the passage of the ova into the tubes, however, the explanation is difficult. At the present time there are two theories in regard to this process : one, in which it is supposed that the fimbriated extremities of the Fallopian tubes, at the time of rupture of the Graafian follicles, become adapted to the surface of the ovaries, and the other, that the ova are carried to the openings of the tubes by ciliary currents. Neither of these theories, however, is susceptible of actual demonstration, and their value is to be judged from anatomical facts.

PUBERTY AND MENSTRUATION.—At a certain period of life, usually between the ages of thirteen and fifteen, the human female undergoes a remarkable change and arrives at what is termed puberty. At this time there is a marked increase in the general development of the body ; the limbs become fuller and more rounded, a growth of hair makes its appearance upon the mons veneris, the mammary glands increase in size and take on a new stage of development, Graafian follicles enlarge, and one or more approach the condition favorable to rupture and discharge of the ova. The female becomes capable of impregnation, and continues so, in the absence of pathological conditions, until the cessation of the menses.

The age of puberty is earlier in warm than in cold climates, and many instances are on record in which the menses have appeared exceptionally much before the usual period. Gen-

erally at the age of forty or forty-five the menstrual flow
becomes irregular, occasionally losing its sanguineous character, and it usually ceases at about the age of fifty years.
In a condition of health the periods recur every month until they cease, at what is termed the change of life. In the
majority of cases the flow recurs on the twenty-seventh or
twenty-eighth day, but sometimes the interval is thirty days.
As a rule, also, utero-gestation, lactation and severe diseases, acute and chronic, suspend the periods, but this has
exceptions, as some females menstruate regularly during
pregnancy, and it is not very uncommon for the menses to
appear during lactation.

Removal of the ovaries, especially when this occurs before
the age of puberty, usually is followed by the arrest of menstruation. It is a well-known fact that animals do not present the phenomena of heat after extirpation of the ovaries.
Cases have been quoted of this operation in the human subject in which the menses were arrested, but this rule does
not appear to be absolute, as other cases have been reported
where menstruation continued with regularity for more than
a year.

When a cow gives birth to twins, one a male, the other
apparently a female, the latter is called a free-martin and has
no ovaries ; it does not breed or show any inclination of the
bull. They are said to possess a rudimentary uterus, but no
ovaries.

The menstrual period presents three stages : first, invasion ;
second, a stage of sanguineous discharge ; third, cessation.

The stage of invasion is variable in different females.
There is usually anterior to the establishment of the flow,
more or less of a feeling of general *malaise*, a sense of fulness
and weight in the pelvic organs, accompanied with a greater
or less increase in the quantity of vaginal mucus, which becomes brownish or rusty in color and has a peculiar odor.

At this time, also, the breasts become slightly enlarged. This stage may continue for one or two days, although in many instances the first evidence of the access of a period is a discharge of blood.

When the symptoms above indicated occur, the general sense of uneasiness usually is relieved by the discharge of blood. During this, the second stage, blood flows from the vagina in variable quantity, and the discharge continues from three to five days. With regard to duration of flow, there are great variations in different individuals. Some women present a flow of blood for only one or two days, while in others the flow continues from five to eight days, within the limits of health. A fair average, perhaps, is four days.

It is rather difficult to arrive at or even approximate the total quantity of the menstrual flow ; it has been estimated at anywhere from four to twelve ounces. It is well known that the quantity is variable, but an average is not far from five or six ounces.

CHAPTER XLVII.

GENERATIVE ORGANS OF THE MALE.

From a physiological stand-point, as regards the anatomy of the male organs of generation, we are to deal mostly with the testicles, in which the male element is produced. In relation with the testicles we have *vas deferentia, vesiculæseminales* and *penis*.

THE TESTICLES.—The essential organs of reproduction in the male are two symmetrical oblong glands, about one inch and a half long, compressed from side to side and situated during a certain period of intrauterine life in the abdominal cavity, but finally descending into the scrotum. Immediately beneath the skin of the scrotum is a loose, reddish, contractile tissue, called the dartos, which forms two distinct sacs, one enveloping each testicle, the inner portion of these sacs fusing in the median line to form the septum. Within this covering the two testicles are distinct and suspended in the scrotum by the two spermatic cords, the left usually hanging a little lower than the right. The coverings for the two testicles, in addition to the ones just mentioned, are the inter-columnar fascia, the cremaster muscles, the infundibuliform fascia, the tunica vaginalis and the proper fibrous coat. The tunica vaginalis is a shut sac of serous membrane covering the testicle and epididymis and reflected from the posterior borders of the testicle to the wall of the scrotum, lining the cavity occupied by the testicle on either side and also extending over the spermatic cord. The spermatic cord is composed of the vas deferens, blood-vessels, lymphatics and nerves, with the covering already described, which expand and surround the testicles.

The proper fibrous coat of the testicles is called the *tunica albuginea*. It is white, dense, inelastic and measures about one-twenty-fifth of an inch in thickness, and is simply for the protection of the contained structures.

Sections of the testicles extend in various directions, show an incomplete, vertical process of the tunica albuginea called the *corpus Highmorianum* or the *mediastinum testis*.

Lining the tunica albuginea and following the mediastinum and the process which penetrate the testicles is a tunic, composed of blood-vessels and delicate, connective tissue called the tunica vasculosa or pia-mater testis. The substance of the testicles is made up of *seminiferous tubules*, which exist to the number of eight hundred and forty in each testicle, and constitute almost the entire substance of each lobule. The entire length of these tubules, when unraveled, is about thirty inches and diameter one one-hundred and seventy-

fifth of an inch. As they pass toward the apices of the lobules they become less convoluted and terminate in from twenty to thirty straight ducts, the *vas recta*, which pass upward
through the *mediastinum* and constitute the *rete testis*. At
the upper part of the mediastinum the tubules unite to form
from nine to thirty small ducts, the *vas afferentia*, which become convoluted and form the *globus major* of the *epididymis*;
the continuation of these tubes downward, behind the testicles and a second convolution constitutes the *body* and *globus
minor*.

THE VAS DEFERENS.—The excretory duct of the testicles
is about two feet in length and extends from the epididymis
to the prostatic portion of the urethra, and is a continuation
of the single tube which forms the body and globus minor of
the epididymis.   It is somewhat tortuous near its origin, and
it becomes larger at the base of the bladder just before it
joins the duct of the seminal vesicle to form the ejaculatory
duct.

THE VESICULÆ SEMINALES.—Attached to the base of the
bladder and situated externally to the vas deferentia are the
two *vesiculæ seminales*. They have an external fibrous coat, a
middle muscular coat and an internal mucous coat, covered
by epithelium, which secretes a mucous viscid fluid. The
vesiculæ seminales serve as reservoirs, in which the seminal
fluid is temporarily stored up. The ejaculatory duct is about
three-fourth of an inch in length, opens into the urethra, and
is formed by the union of vas deferentia and the ducts of the
vesiculæ seminales.

THE PROSTATE GLANDS.—This organ, except as it secretes a fluid which forms a part of the ejaculatory semen,
has chiefly a surgical interest. It is enveloped in a very
dense fibrous coat and contains many glandular structures
opening into the urethra, and presents a great number of non-
striated with a few striated muscular fibres, some just be-

14

neath the fibrous coat and others penetrating its substance and surrounding the glands.

GLANDS OF URETHRA.—In front of the prostates, opening into the bulbous portion of the urethra, are two small racemose glands, called the glands of Cowper. These have each a single excretory duct, are lined throughout with cylindrical epithelium and secrete a viscid, mucus-like fluid, which forms a part of the ejaculated semen. Sometimes there exists only a single gland, and occasionally, though rarely, both are absent. Their uses are probably not of very much importance.

MALE ELEMENTS OF GENERATION.—The *spermatozoids* are the *essential* male element of generation, and these are produced in the substance of the testicles, by a process analogous to that of the development of other true anatomical elements. The testicles can not be regarded strictly as glandular organs. They are analogous to the ovaries, and they are the only organs in which spermatozoids can be developed, as the ovaries are the only organs in which the ovum can be formed. If the testicles be absent the power of fecundation is lost, as none of the fluids secreted by accessory organs of generation is able to perform the office of fecundation.

In the healthy male, at the climax of a normal venereal organism, 11-6 to 92-6 grains of seminal fluid are ejaculated with considerable force from the urethra by an involuntary muscular spasm. This fluid requires about four days for its full restoration. The spermatozoa are peculiar anatomical elements, developed within the seminal tubules, and possess the power of spontanous movement. The spermatozoa consist of a conoidal head and long filamentous tail, which is in continuous and active motion. As long as they remain in the vas deferens they are quiescent, but when free to move in the fluid of the vesiculæ seminales become very active.

Cold arrests the actions of spermatozoa, but they can be restored by the application of dense saline and other solutions.

As to *origin*, spermatozoa appears at the age of puberty, and are then constantly forming until an advanced age. They are developed from the nuclei of large round cells contained in the anterior of the seminal tubules, as many as fifteen to twenty developing in a single cell. When the spermatozoa are introduced into the vagina they pass readily into the uterus and through the Fallopian tubes toward the ovaries, where they remain and retain their vitality for a period of eight to ten days. During the return of the spermatozoa with the ovum, fecundation takes place just outside the womb and in the Fallopian tubes. After floating around the ovum in an active manner, they penetrate the vitelline membrane, pass into the interior of the vitellus, where they lose their vitality, and along with the germinal vesicle entirely disappear.

CHAPTER XLVIII.

FECUNDATION AND DEVELOPMENT OF THE OVUM.

So far as the male is concerned, coitus is rendered possible by the erection of the penis. This may occur before puberty, but at this time intercourse can not be fruitful. Coitus may be impossible in old age, from absence of power of erection ; but spermatozoids may still exist in the generative organs, and fecundation might occur if the seminal fluid could be discharged into the generative passage of the female. Coitus may take place with the female before puberty or after the cessation of the menses, but intercourse can not then be fruitful.

The spermatozoa once within the cervix uteri, and its fusion with the germinal vesicle, there remains a transparent, granular, albuminous substance, in the center of which a new nucleus soon appears. This constitutes the *parental cells*, and is the first stage in the development of the new being. Following this vitellus undergoes *segmentation*. A constriction appears on the opposite side of the vitellus, which gradually deepens until the yelk is divided into two segments, each of which has a distinct nucleus and nucleolus. These two segments undergo a further division into four, the four into eight, the eight into others, and so on until the entire vitellus is divided into a great number of cells, each of which contains a nucleus and nucleolus.

HEREDITARY TRANSMISSION.—As regards the question which determines the sex of offspring, it has been shown that the proportion of male to female births is about 104 to 105, these figures, however, presenting certain modifications under varying conditions of climate, season, nutrition, etc.; but nothing has ever been done in the way of procreating male or female children at will.

No definite rule can be laid down with regard to the transmission of mental or physical peculiarities to offspring. Sometimes the progeny assumes more the character of the male than of the female parent, and sometimes the reverse is the case without any reference to the sex of the child. Sometimes there appear no such relations, and occasionally peculiarities are observed derived apparently from grandparents. This is true with regard to pathological as well as physiological peculiarities, as in the inherited tendencies to certain diseases, malformations, etc.

A peculiar and seemingly inexplicable fact is that previous pregnancies have an influence upon offspring. This is well marked in breeders of animals, but the same influence is frequently observed in the human subject. A woman may have by a second husband children who resemble the former husband. A white woman may have children by a negro man, and subsequently bear children to a white man, these children presenting some of the unmistakable peculiarities of the negro race.

CHANGES IN THE FECUNDATED OVUM.—It is probable that the ovum is fecundated either just as it enters the Fallopian tube, or in the dilated portion, near the ovary. As it passes down the tube it becomes covered with an albuminous coating. This coating, no doubt, serves to protect the fecundated ovum, but finally this albuminous covering liquefies and penetrates the vitelline membrane, furnishing, it is thought, matter for the nourishment and development of the vitellus.

PRIMITIVE TRACE.—Soon after the formation of the *blastodermic vesicle*, at a certain point there is an accumulation of cells, which marks the position of the future embryo, this spot, first circular, then elongated, and forms the *primitive trace*, around which is a clear space, the *area pellucida*, which is itself surrounded by a darker region, the *area opaca*. The primitive trace soon disappears, and the area pellucida becomes elongated ; a new groove, the *medullary grooves*, is now formed, which develop from before backward, and becomes the neural canal.

BLASTODERMIC LAYERS.—The embryo, at this period, consists of three layers, viz : The external and internal blastodermic membranes and a middle membrane formed by a genesis of cells from their internal surfaces. These layers are known as the epiblast, the mesoblast and hypoblast.

The *epiblast* gives rise to the central nervous system, the epidermis of the skin and its appendages and the primitive kidneys.

The *mesoblast* gives rise to the dermis, muscles, bones, nerves, blood vessels, sympathetic nervous system, connective tissue, the urinary and reproductive apparatus and the walls of the alimentary canal.

The *hypoblast* gives rise to the epithelial lining of the alimentary canal and its glandular appendages, the liver and pancreas, and the epithelium of the respiratory tract.

FORMATION OF THE MEMBRANES.—In the mammalia a portion of the *blastoderm* is developed into membranes, by which a communication and union is established between the ovum and the mucous membrane of the uterus. From the ovum two membranes are developed ; one non-vascular, the amnion, and another, the allantois, which is vascular.

The two layers of *decidua* are formed from the mucous membrane of the uterus. At a certain part of the uterus a vascular connection is established between the mucous membrane and the allantois, and the union of these two structures forms the placenta. The fœtal portion of the placenta is connected with the fœtus by the vessels of the umbilical cord, and the maternal portion is connected with the great uterine sinuses.

THE AMNION.—The amnion appears shortly after the embryo begins to develop, and is formed by the folds of the epiblast and the external layer of the mesoblast rising up in front and behind and on each side ; these amniotic folds gradually extend over the back of the embryo to a certain point, where they coalesce and inclose a cavity, the amniotic cavity. Soon after the development of the amnion the allantois is formed ; this membrane is vascular. It encroaches upon and takes the place of the external amniotic membrane, and is covered with hollow villi, which also takes the place of the villi of the amnion, and when this outer coat recedes and becomes blended with the vitelline membrane, it constitutes the *chorion*, the external covering of the embryo.

THE ALLANTOIS.—The allantois is a highly vascular growth, arising from the posterior portion of the peritoneal cavity, which gradually pushes its way out through the amniotic folds, attaching itself to the outer layer of the amnion ; finally, from increased growth, it completely surrounds the embryo, and its edges become fused together.

AMNIOTIC FLUID.—The process of the enlargement of the amnion shows that the amniotic fluid gradually increases in quantity as the development of the fœtus progresses. At term the entire quantity is variable, being rarely more than two pints ; the average is about one pint. It is a clear, transparent fluid containing albumin, glucose, fatty matter, urea and inorganic salts, etc. Its reaction is usually neutral or faintly alkaline, though sometimes it is feebly acid in the latest periods of pregnancy.

THE CHORION.—The external investment of the embryo is formed by a fusion of the original vitelline membrane, the external layer of the amnion, and the allantois. The external surface now becomes covered with villous processes, which increase in number and size by the continued budding and growth of club-shaped processes from the main stem, and gives to the *chorion* a shaggy appearance. They consist of a homogeneous granular matter, and are penetrated by branches of the blood-vessels derived from the aorta.

At about the end of the second month the villosites begin to atrophy and disappear from the surface of the chorion, with the exception of those situated at the points of entrance of the fœtal blood-vessels, which occupy about one-third of its surface, where they continue to grow larger, become more vascular, and ultimately assist in the placenta; the remaining two-thirds of the surface lose its villi and blood-vessels, and become a simple membrane. As the ovum enlarges over a certain area surrounding the point of attachment of the pedicle which connects the chorion with the embryo, the villi are developed more rapidly than over the rest of the surface.

UMBILICAL CORD—Connects the fœtus with that part of the chorion which forms the fœtal side of the placenta. It is a process of the allantois, and connects two arteries and a vein, which have a more or less spiral direction. It appears at the end of the first month, and gradually increases in length, until at the end of gestation it measures about twenty inches. The cord is also surrounded by a process of the amnion. It is usually about the size of the little finger, and will sustain a weight of about twelve pounds.

MEMBRANÆ DECIDUÆ.—In addition to the two membranes connected with the fœtus, there are two membranes formed from the mucous membrane of the uterus, which are derived from the mother, and which serve still further to protect the ovum. The anterior of the uterus is lined by a thin, delicate mucous membrane, in which are imbedded immense numbers of tubules, terminating in blind extremities, the *uterine tubules*. At each period of menstruation the mucous membrane becomes thickened and vascular, which condition, however, disappears after the usual menstrual discharge. When the ovum becomes fecundated the mucous membrane takes on a new growth, becomes more hypertrophied and vascular, sends off little processes or elevations from its surface, and constitute the *decidua vera*.

As development advances, the decidua vera becomes extended, loses its vessels and glands, and is reduced to the condition of a simple membrane, or, rather, *decidua reflexa*, which is thinner than the decidua vera, and grows up on all sides and inclose the ovum, while the villous processes of the chorion insert themselves into the uterine tubules, and in the mucous membrane between them.

THE PLACENTA.—At about the end of the second month the villi of the chorion become enlarged and arborescent over that part which eventually forms the fœtal portion of the placenta. They are then highly vascular and are imbedded in the soft substance of the hypertrophied mucous membrane. At the same time the villi over the rest of the chorion are arrested in their growth, and they finally disappear during the third month. The blood vessels penetrate the villi in the form of loops at about the fourth week, and the placenta is distinctly marked at about the end of the third month. The placenta then rapidly assumes the anatomical characters observed after it may be said to be fully developed.

The fully formed placenta occupies about one-third of the uterine mucous membrane, is generally rounded or ovoid in shape, and of all the embryonic structures, the placenta is the most important. It is seven to nine inches in diameter, six to eight inches in breadth and about one inch thick at the point of penetration by the umbilical cord. Its weight varies from fifteen to thirty ounces. It is most frequently situated at the upper and posterior part of the inner side of the uterus.

The placenta is composed of two portions, a *fœtal* and a *maternal portion*. The *fœtal portion* is formed by villi of the *chorion*, which by developing, rapidly increases in size and number. They become branched and penetrate the uterine tubules, which enlarge and receive their many ramifications. The capillary blood vessels in the anterior of the villi also enlarge and freely anastomose with each other.

The *maternal portion* is formed from that part of the hypertrophied and vascular decidual membrane, between the ovum and the uterus, the *decidua serotina*. As the placenta increases in size the maternal blood-vessels around the tubules become more and more numerous, and gradually fuse together, forming great lakes, which constitute sinuses in the walls of the uterus.

In the human subject, the maternal and fœtal substances are so closely united that they can not be separated from each other, and as the latter period of gestation approaches the villi is extended deeper into the decidua, while the sinuses in the maternal portion become large and extend farther into the chorion. Finally, from excessive development of the blood-vessels, the structure between them disappears, and as their walls come in contact they fuse together so that ultimately the maternal and fœtal blood are only separated by a thin layer of a homogeneous substance. When fully formed the placenta consists principally of blood-vessels interlacing in every direction. The blood from the mother passes from the uterine vessels into the reservoirs (or lakes) surrounding the villi ; but there is not at any time an intermingling of blood, the two being separated by a delicate membrane formed by a fusion of the walls of the blood-vessels and the walls of the villi and uterine sinuses.

The function of the placenta, besides nutrition, is that of a *respiratory organ*, permitting the oxygen of the maternal blood to pass by osmosis through the delicate placental membrane into the blood of the fœtus ; at the same time permitting *carbonic acid* and other waste materials, the result of nutritive changes in the fœtus, to pass into the maternal blood, to be carried off through various eliminating organs. Through the placenta also pass all the nutritive materials of the maternal blood, which are essential for the development of the embryo. In other words, the placenta is to the embryo what the lungs and stomach are to the man.

CHAPTER XLIX.

DEVELOPMENT OF THE EMBRYO.

THE OVUM.—The product of generation retains the name of ovum until the form of the body begins to be apparent, when it takes the name of *embryo*. At the fourth month, about the time of quickening, it takes the name of *fœtus*, a name which it retains during the rest of intra-uterine life.

NERVOUS SYSTEM.—In reading your anatomy on the development and ossifications of the spinal column, how the dorsal or medullary plates close over the groove for the neural canal, and how in the interior of this canal the cerebro-spinal axis is developed by cells which gradually encroach upon its caliber until there remains only the small central canal of the spinal cord, communicating with the ventricles of the brain.

The external surface of this canal gives rise to the *dura mater* and *pia-mater*. The neural canal thus formed is a tubular membrane ; it terminates posteriorly in an oval dilatation, and anteriorly in a bulbous extremity, which soon become partially contracted and forms the anterior, middle, and posterior cerebral vesicles, from which are ultimately developed the cerebrum, the corpora quadrigemina, and medulla oblongata, respectively.

The nerves are not produced as prolongations from the spinal cord into various tissues, nor do they extend from the tissues to the cord, but they are individually developed in each tissue, by a separation of histological elements, from the cells of which the parts are originally constituted. The nerves of the sympathetic system are developed in the same way.

The mode of development of the spinal cord is thus very simple ; but with the growth of the embryon dilatations are observed at the superior and at the inferior extremities of the neural canal. The cord is nearly of uniform size, in the dorsal regions, marked only by the regular enlargements at the sites of origin of the spinal nerves; but there soon appears an ovoid dilatation below, which forms the lumbar enlargement, from which the nerves are given off to the inferior extremities, and the branchial enlargements above, where the nerves of the superior extremities take their origin. The *anterior vesicle* soon subdivides into two secondary vesicles, the larger of which becomes the hemispheres, the smaller the optic thalami : the posterior vesicle also divides into two, the anterior becoming the cerebellum, the posterior the pons Varolii and medulla oblongata.

While this division of the primitive cerebral vesicles is going on, the entire chain of encephalic ganglia becomes curved from behind forward, forming three prominent angles. As development advances the relative size of the encephalic masses change ; the cerebrum developing more rapidly than the posterior portion of the brain, soon grows backward and arches over the optic thalami and the tubercula quadrigemina; the cerebellum overlaps the medulla oblongata. At first the surface of the cerebral hemispheres is smooth, but at about the fourth month it begins to be marked by future fissures and convolutions.

So far as the action of the *nervous system* of the fœtus is

concerned, it is probably restricted mainly to reflex phenomena depending upon the spinal cord, and that perception and
volition hardly exists ; but there is no doubt that many reflex
movements takes place in utero. When a fœtus is removed
from the uterus of animals, even in the early months of pregnancy, movements of respiration occur ; and it is well known
that efforts of respiration sometimes take place within the
uterus. These are due to the want of *oxygen-carrying*
blood in the medulla oblongata when the placental circulation is interrupted.

DEVELOPMENT OF DIGESTIVE APPARATUS.—The intestinal
canal is the first formation of the digestive system. This is
at first open in the greater part of its extent, presenting, at
either extremity of the longitudinal gutter, in front of the
spinal column, a rounded, blind extremity, which is closed
over in front for a short distance. The close of the visceral
plains then extend laterally and from the two extremities
of the intestine, until only the opening remains for the passage of the umbilical cord and the pedicle of the umbilical
vesicle. There is at first an opening communicating between
the lower part of the intestinal tube and the allantois, which
forms the canal known as the urachus ; but that portion of
the communication which remains closed in the abdominal
cavity, becomes separated from the urachus, is dilated and
eventually forms the urinary bladder. When the bladder is
first shut off it communicates with the lower portion of the
intestine, which is called the *cloaca ;* but it finally loses this
connection and presents a special opening, the urethra. As
development advances the intestine grows rapidly in length
and becomes convoluted. It is held loosely to the spinal column by the mesentery, a fold of the peritoneum, this membrane being reflected along the walls of the abdominal cavity. At the upper part of the alimentary canal and anterior
portion, two pouches appear, which become the *cardiac* and

pyloric extremities of the stomach. At about the seventh week the inferior extremity of the intestine is brought into communication with the exterior, by an opening, the *anus*. Anteriorly the mouth and pharynx are formed by an involution of epiblast, which gradually deepens until it communicates with the fore-gut. A short distance below the attachment of the pedicle of the umbilical vesicle to the intestine appears a rounded diverticulum, which is eventually developed into the cæcum, which gradually recedes from the vicinity of the umbilicus, which is its original situation, and finally becomes fixed by a shortening of the mesentery in the right iliac region.

The development of the *anus* is very simple. At first the intestine terminates below in a blind extremity, but about the seventh week a longitudinal slit appears below the external organs of generation, by which the rectum opens. This is the anus. The opening of the anus appears about a week after the opening of the mouth, at or about the seventh week.

THE LIVER—Appears very early and, indeed, at the end of the first month this organ has attained a large size. It first appears as a slight rudimentary protrusion from the side of the alimentary canal. It grows very rapidly, and attains a large size, and almost fills up the abdominal cavity. The hepatic cells are derived from the intestinal epithelium, the vessels and connective tissue from the mesoblast. During the early part of fœtal life the liver is proportionately larger than later, and its weight compared to the weight of the body at different ages is as follows : At the end of the first month, 1 to 3 ; at four to five months, 1 to 6 ; at term, 1 to 18 ; in the adult, 1 to 36. (Burdach.)

THE PANCREAS—Appears at the left side of the duodenum by the formation of two ducts leading from the intestines, which branch and develop glandular structure at their extremities. The *spleen* is developed about the same time, at the greater curvature of the stomach, and becomes distinct during the second month.

The *Lungs* are developed from the anterior portion of the œsophagus. At first a small bud appears which, as it lengthens, divides into branches; secondary and tertiary processes are given off these, which form the bronchial tubes and air cells. The lungs at first extend in the abdominal cavity, but become confined to the thorax by the development of the diaphragm.

THE BLADDER—Is formed by a dilatation of that portion of the allantois remaining within the abdominal cavity. It is at first pear-shaped, and communicates with the intestine, but later becomes separated and opens externally by the urethra. It is attached to the abdominal walls by a rounded cord, the uracus, the remains of a portion of the allantois.

THE FACE.—The anterior portion of the embryon remains open in front long after the medullary plates have met at the back and inclosed the neural canal. The common cavity of the thorax and abdomen is closed by the growth of the visceral plates which meet in front. In the facial and cervical regions the visceral laminæ send up three processes, the *visceral arches*, separated by clefts, the *visceral clefts*.

From the above sketch it is seen that the face and neck are formed by the advance and closure in front of projections from behind, in the same way as the cavities in the thorax and abdomen are formed. The *first*, or the *mandibular arches*, unite in the median line to form the lower jaw, and superiorly from the malleus, but it must be remembered that in the very first stage of development of the head there is no appearance of a face.

A process jutting from the *visceral arches* (or the first formation) grows forward from its base and unites with the fronto-nasal process growing from above, and forms the upper jaw. When the superior maxillary processes fail to unite, there results the cleft-palate deformity ; if the integument also fails to unite, there results the *hare-lip* deformity. The space above the mandibular arch becomes the mouth.

The second *visceral arch* advances and forms the incus and stapes bones, the styloid process and ligaments, and the lesser cornua of the hyoid bone. The cleft between the first and second arches partially close up, but there remains an opening at the side which becomes the Eustachian tube, tympanic cavity, and external auditory meatus. The third and fourth arches develop respectively into the body of the greater cornua of the hyoid bone, and into sides of the neck and the larynx, the arytenoid cartilage having developed first. In front of the larynx and just behind the tongue is a little elevation which is developed into the epiglottis. The openings of the nostrils appear during the second month ; a little elevation, the nose, appears between these openings, and the nasal cavity begins to be distinct from the mouth. The lips are distinct during the third month, the tongue about the seventh week.

THE EYE—Is formed by a little bud projecting from the side of the anterior vesicle. It is at first hollow, but soon becomes filled with nervous matter, forming the *optic nerve* and the *retina ;* the remainder of the cavity is filled with the *vitreous body.* The anterior portion of the pouch becomes invaginated and receives the *crystalline lens*, which is a product of the epiblast, as is also the cornea. The *iris* appears as a circular membrane without a central aperture, about the seventh week. The eyelids are formed between the second and third month.

THE INTERNAL EAR—Is developed from the *auditory vesicle*, budding from the third cerebral vesicle ; the membraneous vestibule appears first, and from it diverticula are given off, which become the semi-circular canals and cochlæ. The cavity of the tympanum, the Eustachian tube, and the external auditory canal are remains of the first branchial cleft ; the cavity of this cleft being subdivided into the tympanum and external auditory meatus by the membrana tympani.

THE SKELETON.—In the development of the embryo, one of the structures observed is the *notochord*. This is situated beneath the neural canal and extends the entire length of the body. It is formed of a cord of simple cells, and marks the situation of the vertebral column, though it is not itself developed into the vertebræ, which grows around it and encroaches upon its substance until it finally disappears. In many mammals the notochord presents a slight enlargement at the cephalic extremity, which extends to the auditory vesicles, and it is considerably reduced in size at the caudal extremity. By the sides of this cord are masses of cells which unite in front of the neural canal and are eventually developed into the *vertebræ*. These are called *protovertebræ* or *somites*.

These cells extend around and encroach upon the *chorda dorsalis*, forming the bodies of the vertebræ. They also extend over the neural canal, closing above, and their processes are called the medullary, or dorsal plates Sometimes the dorsal plates fail to close at a certain point in the spinal column, and this constitutes the malformation known as *spina bifida*. From the sides of the bodies of the vertebræ the various processes are formed, and as the spinal column is developed its lower portion presents a projection beyond the plevis which constitutes a temporary caudal appendage, curved toward the abdomen ; but this no longer projects after the bones of the pelvis are fully developed. At the same time the entire vertebral column is curved toward the abdomen and it is twisted upon its axis, from left to right, so that the anterior face of the pelvis presents

15

a right angle to the upper part of the body; but as the inferior extremities and the pelvis are developed the spine becomes straight. The vertebræ make their appearance first in the middle of the dorsal region, from which point they rapidly extend upward and downward until the spinal column is complete.

At the base of the skull, on either side of the superior prolongation of the chorda dorsalis, are two cartilaginous processes, which are developed into the so-called cranial vertebræ. In this cartilaginous mass three ossific points appear, one behind the other. The posterior point of ossification is for the basilar portion of the occipital bone, the middle point is for the sphenoid, and the anterior point is for the anterior portion of the sphenoid. They are ultimately developed into the basilar process of the occipital bone and the body of the sphenoid.

The entire skeleton is at first either membraneous or cartilaginous. At the beginning of the second month centers of ossification appear in the jaws and clavicle. As development advances, the ossific points in all the future bones extend until ossification is complete. The upper and lower jaws and clavicle ossify about the same time from little bony prominences at about the beginning of the second month of intra-uterine life.

Other ossific points, which gradually extend, are also seen in other parts, the head, ribs, pelvis, scapula, metacarpus and metatarsus, and the phalanges of the fingers and toes. At birth the carpus is cartilaginous, and remains so until the second year.

As ossification progresses, the deposit in the various ossific points gradually encroaches until it reaches the joints, which remain encrusted with the permanent articular cartilage.

While the *skeleton* is thus developed, the muscles are

formed from the outer layer of the mesoblast, and the visceral plates close over the thorax and abdomen in front, leaving an opening for the umbilical cord. The deeper layers of the muscular tissue begin to be distinct at the first of the second month. The skin appears soon after the appearance of the outer muscles, and is very delicate and transparent. At the end of the second month the epidermis may be distinguished. At the third month the nails make their appearance, and the hairs begin to grow about the fifth month.

DEVELOPMENT OF THE GENITO URINARY APPARATUS.— The genital and the urinary organs are developed together, and are both preceded by the appearance of two large, symmetrical structures known as Wolffian bodies or the bodies of Oken. These are sometimes called the false or the *primordial kidneys*. These appear at about the thirteenth day, develop rapidly and on either side of the spinal column, and are so large as to almost fill the cavity of the abdomen. The Wolffian bodies consist of tubules which run transversely and are lined with epithelium. Internally they become invaginated to receive tufts of blood-vessels. Externally they open into a common excretory duct, the *duct* of the *Wolffian body*, which unites with the *duct* of the opposite body, and empties into the intestinal canal at a point opposite the allantois. On the outside of the Wolffian body there appears another duct, the duct of Muller, which also opens into the intestine.

Very soon after the Wolffian bodies have made their appearance, there appear at their inner borders two ovoid bodies, which are finally developed into the testicles for the male or ovaries for the female. The testicles or ovaries are developed at the internal or anterior surface of the Wolffian bodies, first appearing in small, ovoid masses. The *tubes* called the *ducts* of Muller, which begin just above and pass

along the external borders of the Wolffian bodies, at
first open into the intestine near the point of entrance of
the Wolffian ducts. In the female their upper extremity
remains free, except the single fimbria, which is connected
with the ovary. Their inferior extremities unite with each
other, and at their point of union they form the uterus.
When this union is incomplete there is the malformation
known as double uterus, which may be associated with a
double vagina. At about the fiftieth day the Wolffian
bodies and their ducts disappear.

In the female the ovaries pass down no farther than the
pelvic cavity; but the testicles, which are at first in the abdomen of the male, finally descend into the scrotum. As
the testicles descend they carry with them the Wolffian
duct and that portion of the Wolffian body which is permanent, constituting the head of the epididymis; at the same
time a cord appears attached to the lower extremity of the
testicles and extending to the symphysis pubis. This is
called the gubernaculum. It is at first muscular, but the
muscular fibres disappear during the later periods of utero-
gestation. The epididymis and the vas deferens are formed
from the Wolffian body and duct. At about the end of the
seventh month the testicles have reached the internal abdominal ring, and at this time a double tubular process of
peritoneum, covered with a few fibres from the lower portion of the internal oblique muscle of the abdomen, gradually descend into the scrotum ; the testicle following the
process of peritoneum, which latter eventually becomes the
visceral and parietal portion of the tunica vaginalis.

At the eight or ninth month the testicles have reached
the external abdominal ring and then soon descend into the
scrotum. The vas deferens passes from the testicle, along
the base of the bladder, to open into the prostatic portion of
the urethra; and as development advances two saculated di-

verticula from these tubes make their appearance, which are attached to the bladder and constitute the vesiculæ seminales. As the ovaries descend to their permanent situation in the pelvic cavity there appears, attached to the inner extremity of each, a rounded cord, analogous to the gubernaculum testis. A portion of this, connecting the ovary with the uterus, constitutes the ligaments of the ovary, and the inferior portion forms the round ligaments of the uterus, which pass through the inguinal canal and is attached to the symphysis pubis.

Behind the Wolffian bodies, and developed entirely independent of them, the *kidneys*, suprarenal capsules and uterus make their appearance. The kidneys are developed in the form of little rounded bodies, composed of short, blind tubes, all converging toward a single point, which is the hilum.

These tubules increase in length, branch and become convoluted in certain parts of their extent, finally assume the structure and arrangement of renal tubules with their Malpighian bodies, blood-vessels, etc. They all open into the hilum, and during the formation of the kidneys the suprarenal capsules are formed at their superior extremities.

These bodies, the uses of which are unknown, are relatively so much larger in the fœtus than in the adult that they have been supposed to be peculiarly important in intra-uterine life, though nothing definite is known upon the point.

The kidneys are relatively very large in the fœtus. Their proportion to the weight of the body in the fœtus is one to eighty, and in the adult one to two hundred and forty. There is but little doubt that the ureters are developed as tubular processes from the kidneys, which finally extend to open into the bladder.

EXTERNAL ORGANS OF GENERATION—Begin to be de veloped about the fifth week. At the inferior extremity of the body of the embryon a small ovoid eminence appears in the median line, at the lower portion of which there is a longitudinal slit which forms the common opening of the anus and the genital and urinary passages. This is the cloaca. Soon there is formed a septum which divides the rectum from the vagina, the urethra of the female opening above. From the median prominence in the male two lateral rounded bodies make their appearance; these are developed into the glans penis and corpora cavernosa; in the female, from this prominence, are developed the clitoris and the labia minora.

In the male these two lateral prominences unite in the median line and form the spongy portion of the urethra, while in the female there is a want of union and an opening remains between the two labia minora, which is the true vagina.

The scrotum in the male is analogous to the labia majora in the female; the distinction being that the two sides of the scrotum unite in the median line, while the labia majora remains permanently separated. This similarity is further shown by the anatomy of inguinal hernia, in which the intestines descend into the labium in the female and in the scrotum in the male.

THE CIRCULATORY APPARATUS.—The blood and blood vessels develop very early in foetal life, and assume three different forms at different periods of life, all having reference to the manner in which the embryo receives nutritious matter and eliminates products of waste. Vitelline circulation appears first and absorbs nutritious materials from the vitellus. It is formed by blood-vessels which emerge from the body and ramify over a portion of the vitelline membrane composing the *area vasculosa*. The heart lying in the median line, gives off two arches, which unite to form the abdominal aorta, from which two large arteries are given off, passing into the vascular area ; the venous blood is returned by veins which enter the heart. These vessels are known as the *omphalo-mesenteric arteries and veins*. The vitelline circulation is of short duration in the mammals, as the supply of nutritious matter in the vitellus soon becomes exhausted ; however, nutritive matter is absorbed almost exclusively from the mother, by a simple endosmosis, before the placental circulation is established, and by the placental vessels at a later period.

THE PLACENTAL CIRCULATION—Becomes established when the blood-vessels in the allantois enter the *villous processes* of the chorion and come into close relationship with the maternal blood-vessels. This circulation lasts the whole of intra-uterine life, but gives way at birth to the adult circulation, the change being made possible by the development of the true circulatory apparatus.

THE HEART.—The central enlargement of the vascular system in the first circulation, which becomes the heart, is twisted upon itself by a single turn. The portion which is connected with the cephalic extremity of the embryon gives origin to the arterial system, and the portion connected with the caudal extremity receives the blood from the venous system. The walls of the arterial portion of the heart soon become thickened, while the walls of the venous portion comparatively remain thin. There then appears a constriction, which partly separates the auricular from the ventricular portion.

At a certain period of development, the heart presents a single auricle and a single ventricle.

The division of the heart into two ventricles appears before the two auricles are separated. This division is caused by a septum, which gradually extends from the apex of the heart upwards toward the auricular portion; however, at about the end of the second month a septum begins to be formed between the auricles. This extends from the base of the heart toward the ventricles, but it leaves an opening between the two sides—the foramen ovale, or the foramen of Botal—which persists during the whole of fœtal life.

At the anterior edge of the opening of the vena cava ascendens into the right auricle there is a membraneous fold projecting into the auricle. This is called the valve of Eustachius, and it divides the right auricle incompletely into two portions.

During the sixth week the heart is vertical and is situated in the median line, with the aorta arising from the center of its base. At the end of the second month it is raised up by the development of the liver and its point presents forward. During the fourth month it is twisted slightly upon its axis and the point is inclined to the left. At this period the auricular portion is larger than the ventricles, but the auricles diminish in their relative capacity during the latter half of intra-uterine life. The pericardium makes its appearance during the ninth month. In early intra-uterine life the relative size of the heart is very great. The proportion gradually diminishes, until at birth the ratio is nearing the normal. At first it is said to be as one to fifty, at term one to one hundred and twenty, and in the adult one to one hundred and sixty. During about the first half of intra-uterine life the thickness of the two ventricles is nearly the same, but after that time the relative thickness of the left ventricle gradually increases.

FŒTAL CIRCULATION.—Beginning at abdominal aorta, the blood passes into the two primitive iliacus and thence into internal iliacus. From the two internal iliacus the two hypogastric arteries arise, which ascend along the sides of the bladder to its fundus, pass to the umbilicus and go to the placenta, forming the two umbilical arteries. In this way the blood of the fœtus finds its way to the placenta.

The umbilical vein enters the body of the fœtus at the umbilicus; it passes along the margin of the suspensory ligament to the under surface of the liver; it gives off one branch of large size, and one or two smaller ones to the left lobe; it sends a branch each to the lobus quadratus and the lobus Spigelis, and the vessel reaches the transverse fissure. At the transverse fissure it divides into two branches, the larger of which joins the portal vein and enters the liver, and the smaller, which is the ductus venosus, passes to the vena cava ascendens, at the point where it receives the left hepatic vein. Thus the greater part of the blood returned to the fœtus from the placenta passes through the liver, a relatively small quantity being emptied into the vena cava by the ductus venosus.

Containing the placental blood is the vena cava ascendens which pass through the liver, the blood conveyed directly from the umbilical vein by the ductus venosus, and the blood from the lower extremities passes to the right auricle. As the blood enters the right auricle, it is directed by the Eustachian valve, passing behind the valve, through the foramen ovale, into the left auricle. At the same time the blood from the head and superior extremity passes down by the vena cava descendens in front of the Eustachian valve, through the right auricle into the right ventricle. The Eustachian valve is so arranged that the right auricle simply affords a passage for the two currents of blood; the one from the vena cava ascendens through the foramen ovale, pass into the left auricle and left ventricle, and the other from the vena cava descendens, pass through the right auriculo-ventricular opening into the right ventricle.

It is probable, indeed, that there is very little admixture of those two currents of blood in the natural course of fœtal circulation.

The blood passes into the left auricle from the vena cava ascendens through the foramen ovale, passes from the left auricle to left ventricle; the two also receive a small quantity of blood from the lungs by the pulmonary veins. There the left ventricle is filled; at the same time the right ventricle is filled with blood which has passed through the right auricle in front of the Eustachian valve. The two ventricles thus distend, then contract simultaneously. The blood from the right ventricle passes in small quantities to the lungs, the greater part passing through the ductus arteriosus into the descending portion of the arch of the aorta. This duct is about one-half inch in length and about the size of a goose quill. The blood from the left ventricle passes into the aorta and goes to the system.

At birth, the placental circulation gives way to the adult circulation. After a short while the sense of want of air becomes sufficiently intense to give rise to an inspiratory effort and the first inspiration is made. As soon as the child begins to breathe, the lungs expand, blood flows freely through the pulmonary capillaries and the ductus arteriosus begins to contract. The foramen ovale closes about the tenth day. The umbilical vein and ductus venosus become impervious in from three to ten days, and ultimately form rounded cords.

The hypogastric arteries remain pervious at their lower portion, and constitute the superior vesical arteries. A rounded cord, which is the remnant of the umbilical, forms the round ligament of the liver. A slender cord, the remnant of the ductus venosus, is lodged in a fissure of the liver, called the fissure of the ductus venosus.

CHAPTER L.

Points worth Remembering.

Largest Artery—Abdominal aorta.
Largest Nutrient Artery—Tibial.
Largest Synovial Membrane—At the knee joint.
Largest Muscle—Glutæus maximus.
Largest Nerve—Sciaticus magnus.
Largest Vein—Vena cava.
Longest Muscle—Sartorius.
Longest Tendon—Plantaris.

Branchless Artery—Common carotid (except the terminal branches). There are also no branches from the *cervical portion* of the internal carotid.

Veins Carrying Arterial Blood—Pulmonary. (In the fœtus, the veins carrying arterial blood are the umbilical, hepatic and inferior vena cava).

Artery Carrying Venous Blood—Pulmonary. (In fœtus, umbilical, also).

Nerve Perforated by an Artery—Sciatic by the comes nervi ischiadici; the arteria centralis retinæ also pierces the optic nerve.

Nerve Perforated by a Vein—Those just named.

Muscle Perforated by a Muscle—Stylo-hyoid by the digastric.

Muscle Perforated by a Large Nerve—Coraco-brachialis by the musculo-cutaneus.

Vein Perforated by a Nerve—Occasionally the axillary vein by the internal anterior thoracic nerve.

Ligament Perforated by a Nerve—Sacro-sciatic by the anterior branch of the coccygeal nerve.

Ligament Pierced by an Artery—The greater sacro-sciatic, by the coccygeal branch of the sciatic artery. The azygos articularis artery also pierces the posterior ligament of the knee joint.

Membrane Pierced by an Artery—The thyro-hyoid by the superior laryngeal artery.

Tendons Perforated by Tendons—Those of the flexor sublimis digitorum, *of the hands*, for the passage of the tendons of the flexor profundus digitorum. *In the feet*, the tendons of the flexor brevis digitorum are split for the passage of the tendons of the flexor longus digitorum.

Largest Branch of the Internal Carotid Artery—Is the middle cerebral; this is the artery that is liable to become plugged by an embolus.

Bones with no Muscular Attachments—Ten ; ethmoid, nasal, inferior turbinated, vomer, scaphoid, semi-lunar cuneiform, astragalus, middle cuneiform, incus.

Pillars of the Palate—Anterior, formed by projection of palato-glossus muscle ; *posterior*, by projection of palato-pharyngeus muscle.

False Vocal Cords—Formed by *superior* thyro-arytænoid ligaments.

True Vocal Cords—Formed by the *inferior* thyro-arytænoid ligaments.

Hamstrings—Outer formed by tendon of biceps ; *inner*, by the tendons of the gracilis, sartorius, semi-membranosus, and semi-tendinosus.

The Palatal Veins and Muscles, called Azygos—Are *double*, although the term azygos signifies *not paired*.

Veins with Only Epithelial Walls—Those of the diploë.

Amnios was a term given us by Empedocles (B. C. 450). *Aorta* was named by Aristotle (B. C. 384), though he supposed it contained air.

Cataract—The first removal of the lens for this disease was made by Herophilus. Celsus cured the trouble by depressing the lens (couching?).

Dissection—First human dissection after Herophilus' time (Herophilus is said to have dissected 700 subjects) was by Mondini de Luzzi, Professor of Anatomy at Bologna. Old Alexandria, in times before our era, was famous as being the possessor of *two* human skeletons; all Greece and Rome flocked there to see them. Montagana (A. D. 1460) boasted that he had examined *fourteen* human subjects.

Duodenum was named by Herophilus; he also showed the *heart* to be the beginning of *arterial circulation*. In fact, he is the father of anatomy. Fallopius (16th century) said of him, "That to contradict him, was like contradicting the gospels;" that he was "the evangelist of anatomists."

Gynæcologists—The most prominent ones of early date, so far as surgical procedures are concerned, were Paulus Ægineta (early part of the 7th century), though Aetius (close of the 5th century), Galen (A. D. 131), Soranus (A. D. 98–138), Celsus (about A. D. 60), and even Hippocrates (460 B. C.) treated quite lengthily of the subject. Indeed *five* Hippocratic treatises on female troubles were, in early days, in the hands of the medical profession.

Leeches were first employed by Themison (B. C. 30).

Lexicographer, Medical. The first one was Rufus Ephesius, about A. D. 98 or 117.

Lithotomy was extensively practised in old Alexandria, and the famous oath of Hippocrates (460 B. C.) recognized it as undignified for the physician and surgeon.

Nerves—Their functions were discovered by Herophilus; he overthrew the doctrine that they sprang from the brain-membranes, and proved them to come from the brain itself; their *crossing*, near their cranial organ, was first proposed by Aretæus, and he, in this way, accounted for a left-sided head injury resulting in a right-sided paralysis.

Physician—This term was first applied to doctors by the people of Charlemagne, A. D. 805.

Pharmacopœia—The first one was issued by an Arabian, Sabor-Ebr-Sahil (9th century) and was called Krabadin.

Rhinoplasty was devised by Vincent Vianso, an Italian, who lived in the 15th century; also performed by Brauca and Bojani.

Vein valves were first discovered by Fabricius, during the latter portion of the 10th century.

Tricuspid valves of the vena cava were discovered by Erasistratus, a contemporary of Herophilus. He called them *triglochine*.

Torcula Herophili first described (with *Calamus scriptorius*) and named by Herophilus (about 250 B. C.).

Tinctures were first introduced by Arnold, about the year 1315. He was then a professor at Barcelona.

CARDIAC AND PULMONIC CIRCULATION—The venæ cavæ receive the systemic venous blood, and convey it into the right auricle; then it passes into the right ventricle through the tricuspid, or auriculo-ventricular valves, to be thrown into the pulmonic artery (going through the semilunar, or pulmonary valves); is then conveyed to the lungs and oxygenized in the capillary plexus about the intercellular structure and the air-cells, and returned, by the pulmonary veins (4 in number) to the *left* side of the heart into the left auricle; it then passes into the left ventricle (through the mitral valve) to be forced with fifty-one and one-half pounds into the aorta (through the semilunar valves), and from thence to support the system at large.

FŒTAL CIRCULATION.—From the placenta through the umbilical vein to the liver; from thence, by the hepatic veins and *ductus venosus Arantii*, to the inferior vena cava, to the right auricle ; the most of the current, guided by the Eustachian valve, passes through the foramen ovale into the *left* auricle, and from thence into the left ventricle, and from thence into the aorta and system at large. A part of the current, however, enters the *right ventricle*, is then forced into the pulmonary artery, and from the imperviousness of the fœtal lungs is most all conveyed to the aorta by the *ductus arteriosus Botalli.* The blood is at last conducted by the umbilical arteries (branches of the internal iliac) to the placenta for re-oxygenation.

TABLE OF PHYSIOLOGICAL CONSTANTS.*

Mean height of male, 5 feet 6½ inches; female, 5 feet 2 inches.

Mean weight of male, 145 pounds; of female, 121 pounds.

Number of chemical elements in the human body; from 16 to 18.

Number of proximate principles in the human body; about 100.

Amount of water in body weighing 145 pounds; 108 pounds.

Amount of solids in the body weighing 145 pounds; 36 pounds.

Amount of food required daily; 16 ounces meat, 10 ounces bread, 3½ ounces of fat, 52 ounces of water.

Amount of saliva secreted in 24 hours; about 3½ pounds.

Function of saliva ; converts starch into glucose.

Active principle of saliva ; ptyalin.

Amount of gastric juice secreted in 24 hours ; from 8 to 14 pounds.

Functions of gastric juice; converts albumin into albuminose.

*BURBAKER.

Active principles of gastric juice; pepsin and hydrochloric acid.

Duration of digestion; from 3 to 5 hours.

Amount of intestinal juice secreted in 24 hours; about 1 pound.

Function of intestinal juice; converts starch into glucose.

Amount of pancreatic juice secreted in 24 hours; about 1½ pounds.

Active principles of pancreatic juice; trypsin, amylopsin and steapsin. Functions.—1. Emulsifies fats. 2. Converts albumin into albuminose. 3. Converts starch into glucose.

Amount of bile poured into the intestines daily; about 2½ pounds. Functions.—1. Assists in the emulsification of fats. 2. Stimulates the peristalic movements. 3. Prevent putrefactive changes in the food. 4. Promotes the absorption of the fat.

Amount of blood in the body; from 16 to 18 pounds.

Size of red corpuscles; $\frac{1}{3200}$ of an inch.

Size of white corpuscles; $\frac{1}{2500}$ of an inch.

Shape of red corpuscles; circular biconcave disks.

Shape of white corpuscles; globular.

Number of red corpuscles in a cubic millimetre of blood (the cubic $\frac{1}{25}$ of an inch); 5,000,000.

Function of red corpuscles; to carry oxygen from the lungs to the tissues.

Frequency of the heart's pulsations per minute: 72, on the average.

Velocity of the blood movement in the arteries; about 16 inches per second.

Length of time required for the blood to make an entire circuit of the vascular system; about 20 seconds.

Amount of air passing in and out of the lungs at each respiratory act; from 20 to 30 cubic inches.

Amount of air that can be taken into the lungs on a forced inspiration ; 110 cubic inches.

Amount of reserve air in the lungs after an ordinary expiration ; 100 cubic inches.

Amount of residual air always remaining in the lungs ; about 100 cubic inches.

Vital capacity of the lungs; 230 cubic inches.

Entire volume of air passing in and out of the lungs in 24 hours ; about 400 cubic feet.

Composition of the air ; nitrogen, 79.19 ; oxygen, 20.81, per 100 parts.

Amount of oxygen absorbed in 24 hours ; 18 cubic feet.

Amount of carbonic acid exhaled in 24 hours; 14 cubic feet.

Temperature of the human body at the surface ; $98\frac{6}{10}°$ F.

Amount of urine excreted daily ; from 40 to 50 ounces.

Amount of urea excreted daily ; 512 grains.

Specific gravity of urine; from 1.015 to 1.025.

Number of spinal nerves ; 31 pairs.

Number of roots of origin ; two ; 1st, anterior, motor; 2d, posterior, sensory.

Rate of transmission of nerve force ; about 100 feet per second.

Number of cranial nerves ; 12 pairs.

Nerves of special sense.— 1. Olfactory, or 1st pair. 2. Optic, or 2d pair. 3. Auditory, or 8th pair. 4. Chorda tympani for anterior $\frac{2}{3}$ of tongue. 5. Branches of glossopharyngeal, or 8th pair, for posterior $\frac{1}{3}$ of tongue.

Motor nerves to eyeball and accessory structures ; motor oculi, or 3d pair; pathetic, or 4th pair; abducens or 6th pair.

Motor nerves to facial muscles ; portio dura, facial, or 7th pair.

Motor nerve to tongue ; hypoglossal, or 12th pair.

16

Motor nerve to laryngeal muscles; spinal accessory or 11th pair.

Sensory nerve of the face ; tri-facial or 5th pair.

Sensory nerve of the pharynx ; glosso-pharyngeal or 9th pair.

Sensory nerves of the lungs, stomach, etc ; pneumogastric or 10th pair.

Length of spinal cord; 16 or 18 inches, weight 1½ ounces.

Point of decussation of motor fibres; at the medulla oblongata.

Point of decussation of sensory fibres; throughout the spinal cord.

Function of antero-lateral columns of spinal cord; transmit motor impulses from the brain to the muscles.

Functions of the posterior columns; assist in the co-ordination of muscular movements.

Functions of the medulla oblongata; controls the functions of insalivation, mastication, deglutition, respiration, circulation, etc.

Functions of the corpora quadrigemina; physical centers for sight.

Functions of the corpora striata; centers for motion.

Functions of the optic thalami ; centers for sensation.

Function of the cerebellum; center for the co-ordination of muscular movement.

Function of the cerebrum ; center for intelligence, reason and will.

Center or articulate language; 3d frontal convolution on the left side of cerebrum.

Number of coats to the eye; three; 1st, cornea and sclerotic; 2d, choroid ; 3d, retina.

Function of iris; regulates the amount of light entering the eye.

Function of crystalline lens; refracts the rays of light so as to form an image on the retina.

Function of retina; receives the impressions of light.

Function of membrana tympani; receives and transmits waves of sound to internal ear.

Function of Eustachian tube; regulates the passage of air into and from the middle ear.

Function of semicircular canals; assist in maintaining the equipoise of the body.

Function of the cochlea; appreciates the shades and combinations of musical tones.

Size of human ovum; $\frac{1}{125}$ of an inch in diameter.

Size of spermatozoa; $\frac{1}{4000}$ of an inch in length.

Function of the placenta; acts as a respiratory and digestive organ for the foetus.

Duration of pregnancy; 280 days.

TABLE OF ARTICULATIONS, MUSCULAR ATTACHMENTS.
DEVELOPMENT OF CENTERS OF OSSIFICATION.
ETC., IN "OSTEOLOGY."

Name of Bone.	Number of Articulations.	Number of Muscles attached.	Primary Developmental Centers.
Occipital	6	17	4
Parietal	5	1	1
Frontal	12	3	2
Temporal	5	14	4
Sphenoid	12	12	10
Ethmoid	15	none	3
Nasal	4	none	1
Maxillary Sup	9	9	4
Lachrymal	4	1	1
Malar	4	5	1
Palate	7	4	1
Turbinated Inf	4	none	1
Vomer	0	none	2
Maxillary Inf	2	14	2
Hyoid	none	11	5
Sternum	16	10	6
Ribs (12)	24	19	34
Clavicle	3	6	2
Scapula	2	17	7
Humerus	3	24	7
Ulna	2	13	3
Radius	4	9	3
Scaphoid	5	none	1
Semilunar	5	none	1
Cuneiform	3	none	1
Pisiform	1	2	1
Trapezium	4	3	1
Trapezoid	4	1	1
Os Magnum	7	1	1
Unciform	5	2	1
Metacarpal (5)	19	18	10
Ph langes (14)	23	20	28
Vertebræ (24)	72	39	85
Sacrum	4	5	11
Coccyx	1	4	4
Innominatum	3	33	3 and 5
Femur	3	23	5
Patella	1	4	sesamoid
Tibia	3	10	3
Fibula	2	9	3
Calcis	2	8	1
Cuboid	4	1	1
Astragalus	4	none	1
Scaphoid	4	1	1
Int. Cuneiform	4	2	1
M d. Cuneiform		none	1
Ext. Cuneiform	6	2	1
Metatarsal (5)	21	13	10
Ph langes (14)	23	23	28
Malleus	1	3	?
Incus	2	none	?
Stapes	1	1	?

Total in Human Body . . . . 200.

www.ingramcontent.com/pod-product-compliance
Lightning Source LLC
Chambersburg PA
CBHW021527210326

41599CB00012B/1407